Developing Wind Power Projects

Developing Wind Power Projects

Theory and Practice

Tore Wizelius

London • Sterling, VA

First published by Earthscan in the UK and USA in 2007

This publication is adapted from the Swedish-language book *Vindkraft: i teori och praktik* (published by Studentlitteratur, copyright © Tore Wizelius and Studentlitteratur) and has been translated into English by the author. This translation is published by arrangement with Studentlitteratur.

ISBN-13: 978-1-84407-262-0
ISBN-10: 1-84407-262-2

Typeset by Composition and Design Services
Printed and bound in the UK by Cromwell Press, Trowbridge
Cover design by Susanne Harris

For a full list of publications please contact:

Earthscan
8–12 Camden High Street
London, NW1 0JH, UK
Tel: +44 (0)20 7387 8558
Fax: +44 (0)20 7387 8998
Email: earthinfo@earthscan.co.uk
Web: www.earthscan.co.uk

22883 Quicksilver Drive, Sterling, VA 20166-2012, USA

Earthscan is an imprint of James and James (Science Publishers) Ltd and publishes in association with the International Institute for Environment and Development

A catalogue record for this book is available from the British Library

Library of Congress Cataloging-in-Publication Data

Wizelius, Tore.
[Vindkraft i teori och praktik. eng]
Developing wind power projects : theory and practice / Tore Wizelius.
p. cm.
Includes index.
ISBN-13: 978-1-84407-262-0 (pbk.)
ISBN-10: 1-84407-262-2 (pbk.)
1. Wind power. I. Title.
TJ825.W59 2006
333.9'2--dc22

2006013771

The paper used for this book is FSC-certified and totally chlorine-free. FSC (the Forest Stewardship Council) is an international network to promote responsible management of the world's forests.

Mixed Sources
Product group from well-managed forests and other controlled sources
www.fsc.org Cert no. TT-TOC-2082
© 1996 Forest Stewardship Council

Contents

Part V – Wind Power Project Development

List of Boxes, Figures and Tables

Boxes

Figures

Tables

Preface

You don't have to be an engineer to install a wind turbine – there are reliable wind turbines available that are ready to use, and installation is often included in the price. However, to be able to choose the right kind of turbine for the right place, obtain the necessary permissions and get a reasonable economic return on the investment, it is necessary to have a sound knowledge of meteorology, environment laws and economics and a basic understanding of how wind turbines work.

To design and build a wind turbine it is not enough to be an engineer. It takes at least five engineers with different specialities: one mechanical engineer, one electrical engineer, one aeronautical engineer for design of the rotor blades, one computer engineer to design the control system and, finally, a building engineer to design the foundations.

Furthermore, to work with wind power it is not sufficient to learn how a wind turbine operates from a technical point of view. It is just as important to be able to estimate the wind resources at a specific site: the energy content of the wind should always be the basis for a wind power project.

Commercial wind turbines today cost around a million euros or more, so it is necessary to understand the project economics to ensure that the investment will be profitable. It is also necessary to gain the relevant permissions from the appropriate authorities. Therefore it is important to know what impact wind turbines could have on the environment and to be familiar with the laws and regulations that govern the decisions made by the authorities. Finally, the local community should also give their consent. To gain approval for a wind power plant it is vital to inform those who will be affected by the installations and to take their views into serious consideration.

This book aims to provide guidance and the information required to develop commercial wind power plants, through a combination of theory and practice. Laws, rules and project economics will vary in different countries, and change so often that these aspects are described in general terms. The reader should check the specific regulations for the country they are working in.

The Swedish edition of this book, published in 2003, was produced with support from the Swedish Energy Agency, which has contributed to the dissemination of knowledge about wind power not only in Sweden but also internationally. I would like to express my gratitude to Professor Ann-Sofi Smedman at Uppsala University, who has scrutinized the facts in Chapters 4, 'The Wind' and 5, 'The Power of Wind'; to Björn Montgomerie, wind power researcher at FOI (the Swedish defence research agency), who has checked the facts in Chapters 6, 'Conversion of Wind Energy' and 8, 'The Wind Turbine Rotor'; and to my own former teacher Göran Sidén at Halmstad University, who has checked the facts in Chapter 10, 'Electrical and Control Systems'. For this English edition I have updated several facts and figures and the chapters in Part IV, 'Wind Power and Society', have been completely rewritten.

I feel greatly in debt to Paul Gipe, whose eminent books on wind power I have read repeatedly. They have not only been a valuable source of knowledge, but also a great inspiration for my own writing, and I hope that I too have managed to explain things in a comprehensible manner. Gipe's latest book, *Wind Power*, is a masterpiece in both form and content and a book that everyone interested in wind power should read.

I am just as grateful to David Milborrow, who has had the patience to check both facts and the English language in my translated manuscript. I have followed his expert analyses, published in *Windpower Monthly*, for years, and have included some diagrams based on his research in this book. Any faults that may remain in the book are, of course, solely my responsibility.

Tore Wizelius

PART I

Introduction to Wind Power

Wind power is a renewable energy source that has developed rapidly since the end of the 1970s. This has been achieved by an energy policy that has created a market for renewable energy and by research and technical development. In these few decades wind power has developed from an alternative energy source to a new fast-growing industry which no longer needs subsidies and manufactures wind turbines that produce power at competitive costs. This introduction describes this development.

1

Wind Power Today and in the Future

Wind turbines catch kinetic energy in the wind and transform it into other forms of energy: mechanical work in water pumps and windmills or electric power in modern wind turbines. The wind is a renewable energy source; the wind is set in motion by the differences in temperature and air pressure created by the sun's radiation on Earth. Wind turbines produce clean energy, don't need any fuel transport that can be hazardous to the environment, don't create air pollution and don't leave any hazardous waste behind.

The sun, the wind and running water are all renewable energy sources, in contrast to coal, oil and gas, which depend on fossil fuels from mines or oil and gas fields. In many countries, for example Sweden, hydropower has already been fully developed. The technology to use direct solar radiation with solar collectors and photovoltaic (PV) panels is still waiting for a commercial breakthrough that is expected to come during the first decade of this millennium. Wind power is the new renewable energy source that has seen the most successful development so far.

Modern wind turbines are efficient, reliable and produce power at reasonable cost. Furthermore, the wind power industry is growing very fast, with the leading companies having increased their turnover by 30–40 per cent per year in the first years of this decade. Simultaneously the cost per produced kWh has become lower for each new generation of wind turbines that has been introduced on the market.

From the early 1980s the size of wind turbines has doubled approximately every four years. The largest commercial turbines today have hub heights of 110 metres, rotors with a diameter of 110 metres and a rated power of 3.6MW. The next generation, with a rated power of 5MW, have already been built as prototypes. If this development continues, wind turbines may have a rated power of 10MW by 2010.

The technology in the wind turbines has developed in several ways. The control systems have become cheaper and more advanced, new profiles for the rotor blades can extract more power from the wind, and new power electronic equipment makes it possible to use variable speed and to optimize the capacity of the turbines.

Just as wind turbines have grown in size, so installations have also become larger and larger. In the early days of wind power development, turbines were installed one at the time, often next to a farm. After a few years they were installed in groups of 2–5 turbines. Today large wind farms are built, on land and off shore, with the same capacity as a conventional power plant. The largest wind farms in Europe consist of up to one hundred turbines.

A problem with the wind as an energy source is that the wind always varies. When the wind slows or stops, power has to be produced by other power plants. This could lead to the conclusion that it will always be necessary to have a back-up capacity with other power plants with the same capacity as the wind power connected to the power system. If this were true, wind power would be very expensive. However, since wind power only constitutes one part in a large power system, this is not necessary at all. A moderate share of wind power in a system does not need any back-up capacity at all, since it already exists in the power system. In Sweden, for example, power companies can simply save water in the hydropower dams when the wind blows, and use this saved hydropower when the wind drops.

In a power system power consumption varies continually, during each day as well as during seasons. Every power system has a regulating capacity to adapt power production to actual power consumption. This can be used to adapt the system to the variations in the wind – and the output of wind turbines – as well. When the wind power penetration (that is the share of electric power produced by wind in a power system) increases to 10–20 per cent, it may be necessary to regulate the wind power as well, by reducing power from wind turbines in situations with low load (consumption) and high production, or by keeping a power reserve to be used to balance power production with consumption at short notice. Few countries, however, have yet reached such penetration levels.

Fast market growth

During the development from small single turbines connected to farms to large wind farms with the capacity of large-scale power plants, wind power has become more competitive: the power produced by wind turbines has grown cheaper. Today the cost of power produced by wind turbines (in places with good wind conditions) is competitive with the cost of power produced by oil, coal, gas or nuclear fuel in *new* power plants. Within this decade wind power could become the cheapest energy source available.

To lower the cost of wind power still further takes mass production of turbines. To attain this, the market has to grow. And indeed it is growing very fast. Germany, Denmark, Spain and the UK, as well as India and China and the US, are installing wind power plants on a large scale.

During the last few years the German market has grown fastest. Germany passed pioneering Denmark in 1994 with respect to installed wind power capacity and currently has the most wind power installed in the world. German manufacturers are now competing with the Danish industry. Both Germans and Danes

BOX 1.1 WIND POWER STATISTICS

To indicate how much wind power there is in a country, the total installed capacity is used as a measure. Every wind turbine has a rated power (maximum power) that can vary from a few hundred watts to 5000kW (5MW). The number of turbines does not give any information on how much wind power they can produce. How much a wind turbine can produce depends not only on its rated power, but also on the wind conditions. To get an indication of how much a certain amount of installed (rated) power will produce per year, this simple rule of thumb can be used: 1MW wind power produces 2GWh/a on land and 3GWh/a offshore.

1TWh (terawatt hour) = 1000GWh (gigawatt hours)
1GWh = 1000MWh (megawatt hours)
1MWh = 1000kWh (kilowatt hours)
1kWh = 1000Wh (watt hours)

International information on wind turbine installations is available at *www.windpower-monthly.com* (the Windicator), *www.ewea.org* and *www.ieawind.org.*

have found new large export markets in India and China. Spain has installed several thousand MW in the last few years, and in the US large wind farms are installed on the large plains in the Midwest and on the west coast. In Denmark and Germany development on land has now reached a level where it is harder to find new sites for wind turbines and consequently growth in their domestic markets

Table 1.1 *Global wind power capacity in 2005 (MW)*

Country	Installed in 2005	Total 2005
Germany	1798	18,247
Spain	1764	10,027
US	2424	9124
India	1430	4430
Denmark	4	3128
Italy	452	1717
UK	465	1353
China	496	1260
Netherlands	141	1219
Japan	144	1040
Other	2192	7436
Total	**11,310**	**58,981**

In 2005, 11,310MW of new wind power was connected to the world's power grids. Total installed power increased to almost 60GW, an increase of 24 per cent from 2004.

Source: World Wind Energy Association (2006)

has declined. Thus in 2004 Spain took the lead in terms of most installed capacity in a year. In short, most of the growth so far has been concentrated in a few countries and there is still an immense potential for market growth in countries where development has hardly taken off – Australia, Brazil, Ireland, Canada, Poland, Norway, to mention a few.

There are also ambitious plans to develop wind power plants offshore. Several offshore wind farms are already installed in Denmark, the UK, the Netherlands and Sweden. Denmark has decided that wind power shall produce 50 per cent of the electric power in the country by 2030, a significant increase from the 20 per cent today, and the development of large offshore wind farms necessary to realize this ambitious target has already started. The UK has also started an ambitious plan for offshore development.

2
Historical Background

The ancient Greeks had no windmills, and while they used sails on their ships to harness the power of the wind, their knowledge of wind power generally was weak. The Greeks, the Romans and the Vikings used square sails, and steered using oars instead of rudders, which made it hard to keep a straight course when the wind came from the side and avoid drifting away in the direction of the wind. Therefore their ships had large crews – they needed many strong oarsmen and galley slaves to reach a destination within a reasonable time. When winds increased to storm force, the sails were taken down, a simple method of *power control.* In other words they had respect for the unpredictable and unreliable wind that could turn even the best of ships into disabled wrecks.

Exactly how long man has known how to utilize wind for work is unknown, but some kinds of windmills were probably used in China and Japan some 3000 years ago. The first historically well-documented windmill dates from AD947, in Persia, close to the border with Afghanistan. There, as in many other places on Earth, the wind varies according to a regular pattern. Some times of the year the wind always blows from the sea inland, through a pass. Those who built the windmill didn't have to worry about the wind direction.

This windmill had a vertical axis, on which mats were mounted. This is the same principle as for a small watermill. To make it rotate, half of the rotor has to be protected from the wind. This was easily done, since the wind always came from the same direction. A wall surrounded the mill, with one opening facing the wind, so that the wind only hit the mats on one side of the axis. Power control was simple. If the winds blew too hard, a door was closed. At the back there was no wall at all, to utilize the power of wind the air must be able to get away behind the turbine (see Figure 2.1).

By the end of the 12th century the first windmills in Europe had been built on the Mediterranean coast and in northern France. These demonstrate a radical change of technology: while the Persian windmills had a vertical axis, these had a horizontal one.

To be able to build a horizontal axis windmill, another fundamental technological item has to be used – the cogwheel. The power has to be transformed from the horizontal to the vertical axis for the millstones to be turned around. The cog-

Figure 2.1 *Persian windmill*

The Persian vertical axis windmill is sited inside a building where a wall screens off the part of the rotor that turns into the wind.

Source: Bo Göran Johansson

wheel had, however, already been invented and was used in watermills. To transfer this technology to windmills was a simple matter.

With a horizontal axis and a vertical rotor, a new technical problem has to be solved – how to direct the rotor towards the wind. The windmills used in the Mediterranean area, which are still used on Crete, for example, were built with a stone tower. The rotor and the shaft were made of wood, and the wind was caught by sails made of cloth (see Figure 2.2). The rotor was made with quite thin wooden bars and had a low weight. The top of the tower was constructed like a 'cap' and could be turned.

The windmills in northern Europe, in France, the Netherlands and the UK, were at this time so-called post mills. The whole mill was mounted on a strong stand, with a vertical wooden log, so that the whole mill could be turned and the rotor directed towards the wind. It was heavy work. A man could turn the mill using a lever, but if it was too heavy, an ox was used to turn the mill into the wind.

Windmills started their march into Europe in the 13th century and soon became one of the most important power sources, a position maintained until the end of the 19th century. By the middle of the 19th century, when the number of mills reached its peak, there were some 9000 windmills in the Netherlands,

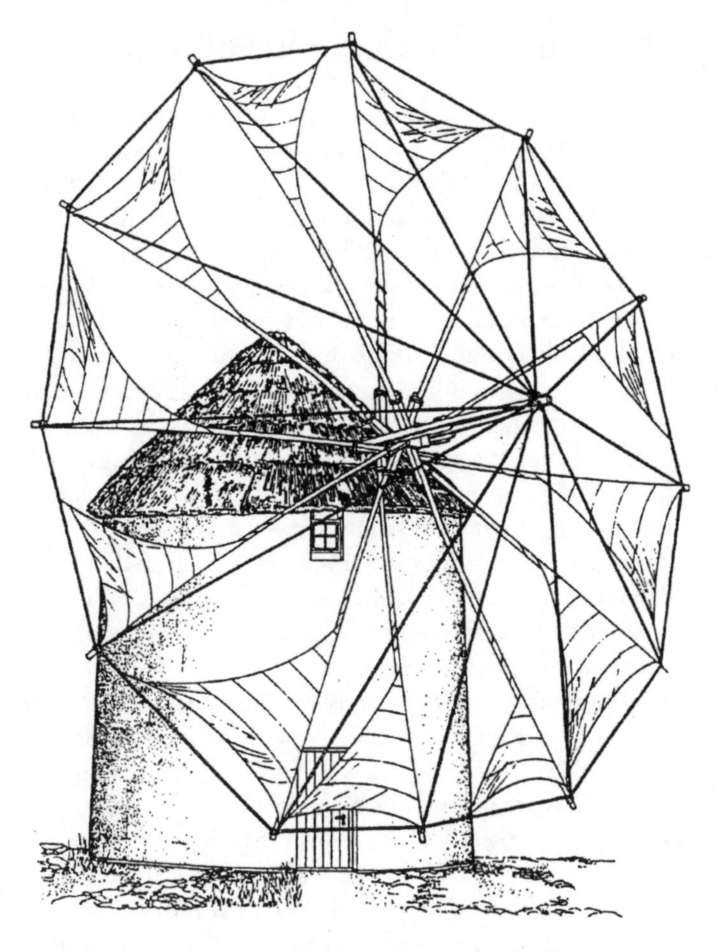

Figure 2.2 *Mediterranean windmill*

To transmit power from a water wheel to a millstone, you need cogwheels. The same technology is used in horizontal axis windmills. The type used in the Mediterranean area has a revolving cap and a rotor with sails.

Source: Hills (1996)

18,000 in Germany, 8000 in England, 3000 in Denmark and 20,000 in France. Those who have read Cervantes' novel *Don Quixote* know that there were also plenty of windmills in Spain. Before the steam engine was developed and modern industrialism took off, hydropower and wind power produced half of the power each, excluding horses and oxen.

The windmill as a symbol of freedom

During these centuries the windmill also played an important political role. 'The windmill was a tool for social development, the number of skilled mechanics increased, and it lessened the burden of women, they no longer had to mill their grain by hand,' writes British historian Edward Kealey, arguing that the windmill undermined feudal oppression and contributed to women's liberation.

During the 14th century the squires and the monasteries had exclusive rights to exploit and use watersheds, and in that way had a watermill monopoly. And they used this monopoly position to exploit the farmers. The wind, however, belonged to all, and the power in the wind was free. Farmers and artisans built windmills, and thus liberated themselves from one of the burdens of the feudal society. These times were characterized by social unrest. For the growing liberation movement, the windmill became both a tool and a symbol of the idea of freedom that spread and threatened the foundations of medieval society.

During the following centuries the simple stock mill underwent impressive technological development. The Dutch, especially, were masters of windmill engineering. They developed a brand new concept that utilized the same principle as the Mediterranean windmill, with a revolving 'cap'.

The windmills also grew in size. Mills with 20–30 metre brick towers and 20–30 metre rotor diameters were built; these produced some 25–30kW. The total power produced by all the windmills in Europe at the peak of the windmill era was 1500MW, a level that wind power did not reach again until 1988.

The windmills in northern Europe usually had four rotor blades, with the outer part built as an 'espalier'. When the mill was about to be used, the blades were covered with wooden plates or with sailcloth. When the wind got too strong, reducing the cover on the blades could regulate power. There were also advanced versions with adjustable 'Venetian blinds'. The windmills were not in use continually but were started when the miller was about to start milling grain and when there was suitable weather.

The windmill engineers also developed an automatic yaw mechanism, the so-called Dutchman. This was made of a wind wheel (or two, one on each side of the tower) that was mounted perpendicular to the rotor. The wind wheel was connected to a cogwheel on the tower by a gear drive. When the wind direction changed so that the rotor was hit from a side angle, this wind wheel was activated and started to turn. The wind wheel turned the 'cap' of the windmill until the rotor again was directed towards the wind. In that position no wind could hit the wind wheel and it stopped turning (see Figure 2.3).

Windmills were not only used to grind grain, although that was the most important application. Another important task they were used for was to pump water, thus contributing to land reclamation for agriculture. The rotor turned either a blade wheel or a water screw. Such wind driven pumps made it possible to increase the land area of the Netherlands. A large part of the country today

Figure 2.3 *The Dutchman*

Windmill technology developed over centuries. The simple stock mill that was used by farmers was further developed into the Dutch mill, which spread through Europe in different versions. The Dutch mill was a tower with a revolving cap and was not used only to grind grain, but also as a sawmill, rag mill and snuff mill and for flax dressing and other mechanical work. This technically advanced windmill – a Dutchman in Britain, with a wind wheel for turning the cap and rotor blades with Venetian blinds for automatic power regulation – is sited in North Leverton and is still in a usable condition.

Source: Hills (1996)

consists of drained seabed, so-called polders. There were also windmills that were used to saw timber, to stamp rag that was used to make paper and for many other applications. In the industrial area outside Amsterdam there were 700 windmills in the 19th century providing mechanical power to the factories.

Steam engines finally drove the windmills out of the market, but it was a surprisingly slow process, in fact taking a whole generation. Nevertheless, in the 20th century the windmill definitely had had its day and its role is now reduced to one of heritage. In the Netherlands, where the windmill has become an important national symbol, the authorities have set a goal to keep at least 1000 windmills in good shape and many of them are actually set to work a couple of days each year.

The wind wheel of the Wild West

On the other side of the Atlantic Ocean, on the American prairie, another kind of windmill played an important role: the wind pump, a wind wheel that was used for water pumping. According to the American historian Walter Prescott Webb there were three inventions that made it possible for man to colonize the prairie: the revolver, barbed wire and the wind pump. A common saying was: 'Women who can't fire a gun or climb a wind pump have no future here.' The prairie was described as a land where 'the wind pumps the water and the cows chop the wood' (since there was no wood dried cow dung was used instead of firewood).

The wind pump was mounted on a post or a truss tower. The pump was driven directly by an axle that was made to rotate by the wind. In the beginning, both the towers and the rotor blades were made of wood; only the axle and some other mechanical parts were made of iron. The pumps were installed next to the farms. They had to be oiled and greased regularly and when the winds increased they had to be turned out of the wind and a brake had to applied manually. They needed constant surveillance.

In the Midwest industrialization took off in the 19th century and in 1854 a wind pump with automatic power regulation, developed by Daniel Halladay, was introduced on the market. It had four twistable rotor blades and when the wind speed increased the force of the wind twisted the blades so that some of the wind passed through the rotor. The rotor was directed towards the wind by a wind vane. This wind pump could work without constant surveillance and could therefore be used to pump water for cattle far from the farmstead, which made it possible for farmers to increase their grazing land and increase the number of cattle (see Figures 2.4 and 2.5).

The market for wind pumps of different sizes grew fast. By the end of the 19th century there were 77 companies that manufactured these types of wind turbines in the US. And simultaneously technical development continued. Halladay's wind pump met market competition from the Eclipse model, which was simpler

Figure 2.4 *Wind wheel as water pump*

This wind wheel from Aermotor was a result of the experiments conducted by the American engineer Thomas Perry, who had tested 5000 different rotor options. The rotor is almost solid (it covers almost the whole swept area) and the blades are slightly cup-shaped.

Source: Gipe (1993)

and had fewer moving parts. This model was, in turn, outshone by Aermotor, developed by the engineer Thomas Perry. Perry performed a large number of experiments to develop an efficient rotor, testing some 5000 different rotor options! His wind pump used a wind wheel, an almost solid rotor with a large number of inclined blades. The efficiency of this rotor was twice as high as competing models and could be manufactured at 20 per cent of the price.

During the 19th century the US railroad network developed fast and railroad tracks where built across the prairie. The steam locomotives needed to fill their water tanks at regular intervals, so next to railway stations out on the prairie very large wind wheels were built, up to 18 metres in diameter, to pump water.

Wind turbines for water pumping, usually wind wheels with around 2.5m rotor diameter, are still used in many parts of the world. In the 1990s there were around 600,000 wind pumps in use in Argentina, some 250,000 in Australia, 100,000 in South Africa and 60,000 on the American prairie. Today there are more than one million such wind-driven water pumps in use in the world.

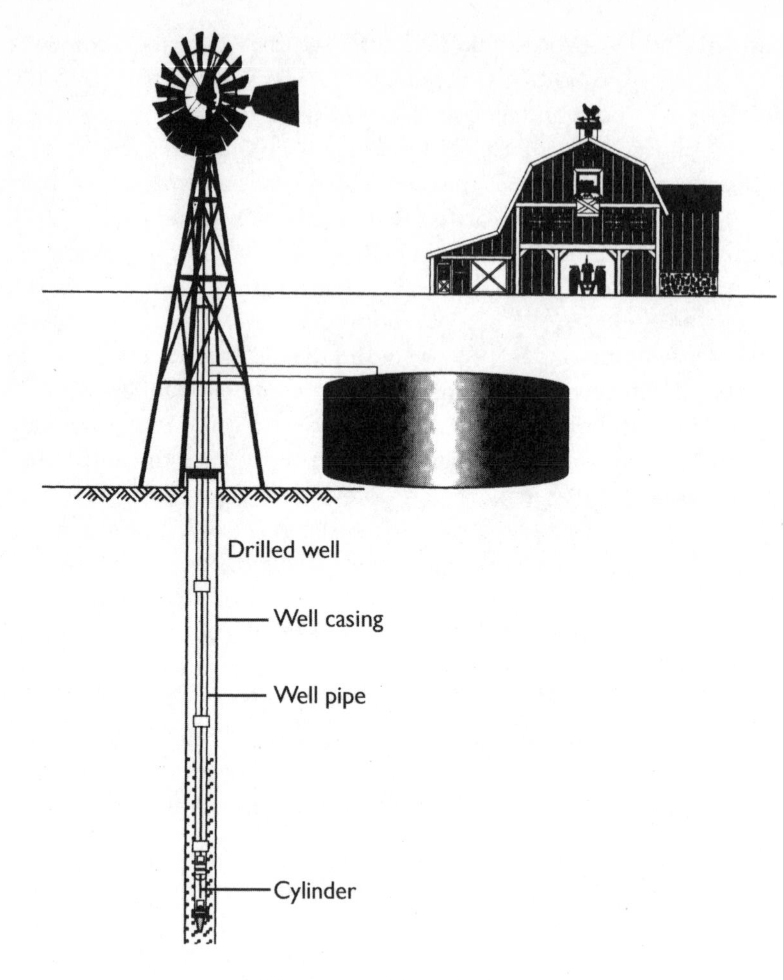

Figure 2.5 *The wind pump*

The wind pumps played an important role in the colonization of the Wild West.

Source: Gipe (1993)

Wind charger

In the 1930s a market for a completely different kind of wind turbine was developed. Electric power had made its entry, but in the early days of electrification, there were electric power grids only in the cities, where the generators were driven by coal, and as small 'islands' around hydropower stations. Large parts of the countryside were without electricity and there was not much demand for it either. In the cities, however, more and more electric appliances were being developed, radios being an example. In the countryside these used crystal receivers, which only needed a small battery, but when the 'real' radio was developed, even people in the countryside wanted to listen to it, and to do that they needed electricity.

To supply farmers in the countryside with electricity a new type of wind turbine was developed – the wind charger, which charged batteries. The aircraft industry had been established for some decades, and the manufacturers of these wind chargers utilized the new knowledge of aerodynamics to give the rotor blades suitable profiles so that the small two- or four-bladed turbines worked like aircraft propellers, only the other way around. These so-called fast runners, with a rotor tip speed five to ten times faster than the wind speed, had a sufficiently high rotational speed to drive a generator without a gearbox. The rotor was connected to a small generator and the current was fed to batteries, which in turn were used to supply radios and light bulbs with electricity. Without battery back-up the radio would be silent during calm days.

The radio manufacturer Zenith and the wind turbine manufacturer Wincharger joined in a marketing campaign to spread these new commodities across the countryside. There were hundreds of thousands of wind chargers operating in the US and several thousands in countries like Denmark and Sweden.

When the electric power grid was finally developed in the countryside the market for wind chargers died. This kind of small wind turbine for electric power generation and battery charging is still manufactured, and has a niche market for different kinds of off-grid applications, being used on oil platforms, in light houses, for radio transmitters and in research stations in the arctic, and nowadays also on sailing boats and caravans used for leisure (see Figure 2.6).

Grid-connected wind turbines

The Danes have always been early starters when it comes to wind power. In 1892, with financial support from the state, Professor Paul la Cour built the first wind turbine for electric power production. It produced DC power and used batteries for energy storage, just like the American wind chargers. The Danish turbines, however, were much larger in size. In 1908 there were 72 wind turbines of 10–20kW online and at the end of the Second World War there were 18 turbines with a nominal power of 45kW. After the Second World War the Danes also tried to connect wind turbines to the power grid.

A turbine with 200kW nominal power and an AC generator was built in Gedser and was put online in 1957 (see Figure 2.7). Simultaneously a wind measurement programme was started to survey the wind resources in Denmark.

In Vermont in the US a gigantic prototype was built in the 1940s, dubbed Grandpa's Knob. This turbine had a rotor diameter of 53 metres and a nominal power of 1250kW. It was online for 1100 hours and fed power to the grid.

Figure 2.6 *Battery charger*

Excenter, an American farm windmill for battery charging, was also sold in Europe. One of these was used at a farm on Gotland, Sweden for several decades. Now it stands as a statue in front of Gotland University.

Source: Tore Wizelius

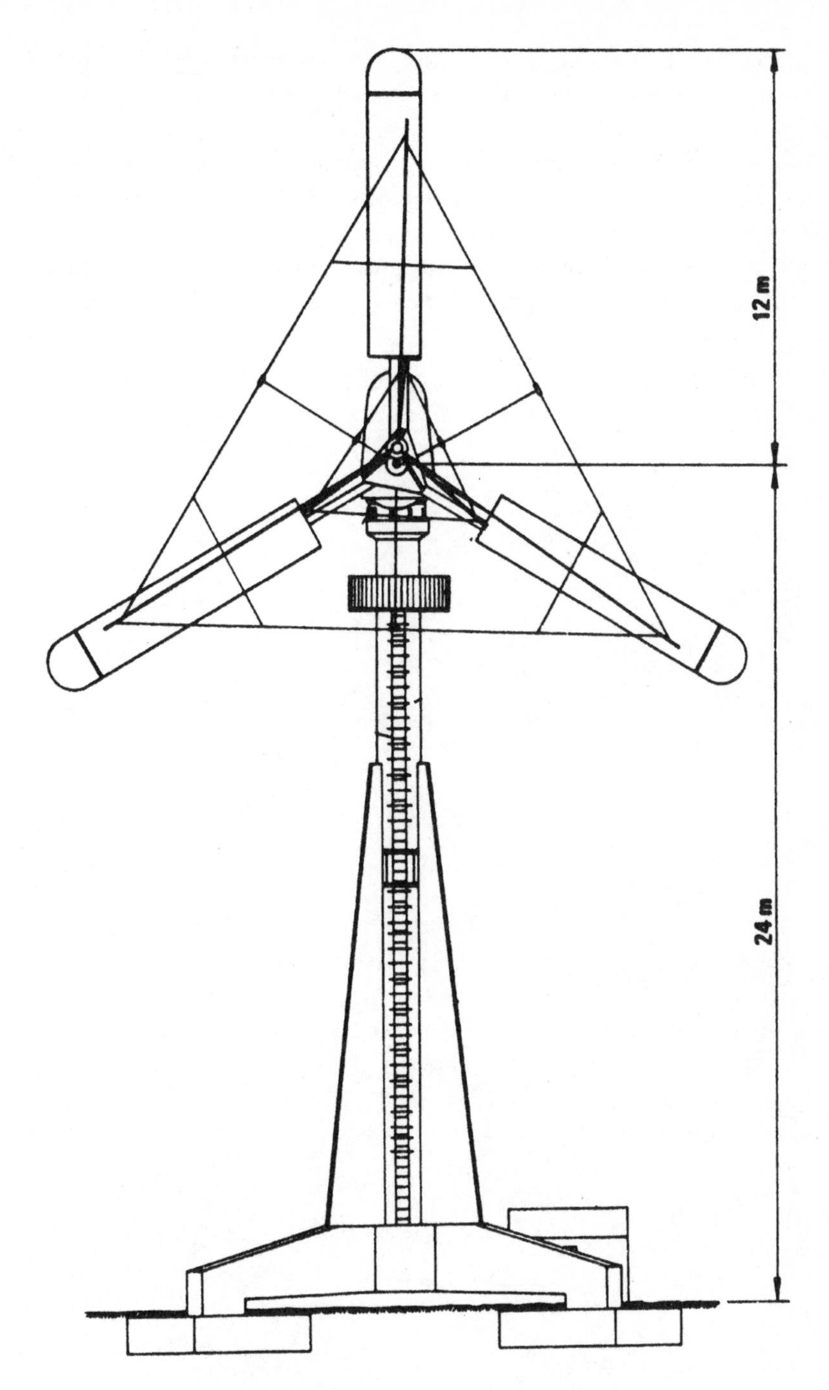

Figure 2.7 *Grid-connected prototype*

The wind turbine in Gedser, Denmark, with 200kW nominal power, was in use from 1957 to 1967.

Source: Södergård (1990)

During the first half of the 20th century two new types of wind turbines were invented: the *Savonius rotor* and the *Darrieus turbine* (see Figures 7.3–7.5, page 76).

A serious effort to develop wind power started at the end of the 1970s, when the oil crises pushed several countries to opt for new energy sources. In the 1980s a new industrial branch was born and developed very fast – the wind power industry. Today this new branch has turned into a competitive industrial branch for large-scale utilization of wind power for energy supply on a global scale.

3
Development of the Modern Wind Power Industry

The oil crises in the 1970s created panic in the industrialized world. This was heightened in the 1980s by the nuclear reactor accidents in Harrisburg and Chernobyl. Both politicians and the public started looking for new energy sources. The option that appeared was to develop local renewable energy sources – solar energy, biomass fuels, hydropower and wind power. During the same period threats to the environment became a focus of public concern and found a place on the political agenda. There was a lively debate about the limitations of the Earth's natural resources, air pollution that crossed national borders and caused acidification and eutrophication, and about persistent chemical compounds like DDT and PCBs. In the 1990s the focus shifted to the ever-increasing emissions of carbon dioxide and other greenhouse gases that threatened to change the global climate.

The burning of fossil fuels for power production was the main source of these emissions; energy policy and environment policy turned out to have a close connection. These sources of damage to the environment were causing great costs for society, through reduced harvests in the agricultural sector, dying forests and increased expenses for health care; these are now referred to as the *external costs* of energy production. At the UN environment conference in Rio de Janeiro in 1992 and in international negotiations about the global climate, aimed at reducing emissions of carbon dioxide and other greenhouse gases, the necessity for change to renewable energy sources has been given ever-stronger weight.

National wind power programmes

By the end of the 1970s most politicians agreed that the dependence on oil had to decrease fast and that it was necessary to develop new and renewable energy sources. The only disagreements concerned how this should be done, who was to be in charge, and how long a time it would take.

The first question the politicians raised was 'What is the best way to develop efficient wind turbines?' Their answer was: 'Invest money in research and development.'

The next question was who could do this. The answer seemed close at hand: power companies and the aircraft industry. The solution involved the state investing large amounts on research and development (R&D), with the task to develop wind power given to large power companies and large industrial manufacturers, since they were most likely to have the competence and the resources necessary to succeed.

Sweden, Germany, the UK, the Netherlands, the US and many other countries started national wind power research programmes according to this model. The aim of all these programmes was to develop large wind turbines, with a nominal power of several MW, which was considered necessary for wind power to be able to make a significant contribution to power production at a national level.

During the 1980s a number of very large prototype turbines were erected in several countries. In Sweden, for example, large turbines with a nominal power of 2–3 MW, Näsudden I and Maglarp, were built. In the US turbines of this size were built too, among others by the aircraft manufacturer Boeing. The Germans built even larger turbines; unfortunately they never managed to get them into regular operation, although they had invested hundreds of millions of marks in the project.

Danes make big effort on small turbines

The Danes chose a different strategy. They chose to create a market for wind turbines. The politicians introduced generous investment grants (subsidies) for wind turbines and a law whereby the state guaranteed a good and reliable price for the wind-generated electric power. This was an assurance to investors that they would get their money back.

This was a wholehearted effort: all the political parties in the Danish parliament supported it. The Danes also invested in R&D as much as other countries, but this money was spent on a wind power research and test centre, Risoe in Roskilde.

In Denmark a completely different kind of company started to manufacture wind turbines. These were farming equipment companies, producing ploughs and so forth. Companies that manufactured ships and machinery for the fishing industry also entered the wind turbine business, and a company that manufactured plastic boats, LM Glasfiber, started to manufacture rotor blades for wind turbines. These were all small and medium-sized enterprises, workshops and smithies. They started out with the knowledge that was available (in Denmark there had already been several prototypes tested in the beginning of the last century and also after the Second World War) and started to produce quite small turbines, of 30–50kW, suitable for farms, for the market.

What farmers demand

What kinds of demands do farmers put on their machines? They should be reliable and robust, stand up to all kinds of weather conditions, and if they break it should be possible (and simple) to repair them in the farm workshop. The Danish companies manufactured wind turbines according to these principles. They did not opt for high-tech but for simplicity and durability. They manufactured turbines of moderate sizes, suitable for farms. The first models were 15–20 metres high and with a nominal power of 20–30kW. The manufacturers had a large and faithful group of customers, a solid market; wind turbines became just another machine among their products.

Many farmers were interested in buying their own wind turbines, even though it was a new technology with apparent economic risks. But there was also something to gain. If they worked well, they would give a reasonable pay-off. And it was not only farmers who were interested: other Danes, aware of the environmental crisis, invested in wind turbines. They formed wind power cooperatives, where a number of persons (households) invested a few thousand Danish kroner each and bought a turbine together so that they could get clean green electric power for their households. Today some 70,000 Danes are part owners of wind turbines or wind farms. The largest wind power cooperative so far was formed in 2000 in Copenhagen, with more than 8000 households collectively owning 10 of the 20 turbines on the Middelgrund offshore wind farm.

The Danes also placed an emphasis on the development of the wind power research centre at Risoe that would test and certify turbines. This created fruitful cooperation and feedback between manufacturers and researchers, who had many good ideas about how the turbines could be developed further. Simultaneously there was an increasing number of turbines online, providing feedback from practical experience, which turned out to be a very valuable asset. The size of the turbines also increased slowly, but at regular intervals, with the nominal power of the turbines on the market doubling every third year.

In the early 1980s the turbines had a nominal power of about 20kW, by 2000 the rating was approaching 2MW. In other words in 20 years the nominal power of the turbines increased by a factor of 100! The next generation to enter the market will be turbines of 5MW; several prototypes are already online. The strategy to create a market for commercial turbines by investment grants and other subsidies in the 1980s turned out to be a much more successful strategy than large investments in high-tech R&D projects.

The wind rush in California

In California a support programme for wind power development and to create a commercial market was introduced in the 1980s through tax credits and favourable power purchase contracts. This created a wind rush, just like the gold rush in an earlier century. A lot of different companies entered the wind power business and started to manufacture turbines with a nominal power of 50–100kW. Some companies scaled up the once-popular wind chargers (stand-alone turbines with battery storage for off-grid operation on remote farms) and adapted them for grid connection. Others focused on newly developed lightweight high-tech turbines. Thousands of wind turbines of this size were installed in a few years, in Tehachapi, Altamont Valley and other windy sites in California, and suddenly the US became the country with most wind power online in the world.

This rapid development of a new industrial branch had an important and fatal drawback, however: the turbines were not reliable. The winds blew them apart and turned the wind farms into graveyards of iron scrap. Thus investors started to look for more robust turbines, and these they found in Denmark. This was the start of the real success story for the Danish wind power industry.

The Californian wind rush came to an abrupt end when the support for wind power was withdrawn in 1986. Many Danish manufacturing companies entered an economic crisis, and several bankruptcies and mergers followed. The domestic market was, however, still alive, and in the beginning of the 1990s Germany introduced its own market stimulation programme, with investment subsidies and guaranteed power purchase prices, just like the one Denmark had introduced a decade earlier. In the same period the investment subsidies in Denmark were withdrawn, since wind power had now become mature enough to stand on its own feet.

Germany and the US start to chase Denmark

When Germany introduced a 100MW programme for wind power development, which was soon extended to a 250MW programme, the Danes got a new fast-growing export market. However, a large share of the support went to German manufacturers, which started to compete with the Danes. In both Germany and Denmark economic support for wind power was seen not only as a part of energy and environment policy but also as part of industrial policy: the support would make manufacturing companies grow, creating new jobs and economic growth.

In the US the market recovered a few years after the withdrawal of support in 1986, and some domestic companies developed wind turbines that could compete with the Danish models. The UK, the Netherlands and Spain also built some wind power plants and developed domestic manufacture of turbines.

Since 2000 Spain has seen tremendous growth in wind power, and the Spanish power companies play an active role in this development. In Spain the political will to develop wind power on a large scale is not only a part of energy policy but also of industrial and regional policy. All turbines that are installed in Spain have to be built in Spain as well, so most large manufacturers have built factories in Spain in order to take part in the market. Many disused shipyards on the north-western Atlantic coast have been taken over by wind power companies and a lot of new jobs have been created in this once depressed region of the country.

In Asia India has played a prominent role as a pioneer for wind power. In the 1990s several European manufacturers established joint ventures there, and some domestic companies also began to manufacture wind turbines. Today one of the most successful Indian companies, Suzlon, has started to opt for the export market and to compete with other manufacturers in Europe and the US. In China there has been some development as well, and in Japan the pace of wind power development has increased in recent years, with the involvement of domestic manufacturers like Mitsubishi. Wind power is now utilized on a small or large scale on all continents and in most countries in the world.

This rapid market growth, however, has so far been concentrated in just a few countries, like Denmark, Germany, Spain, India and the US. In the coming years it is expected that other countries – Canada, Australia, Brazil and the countries in Europe where development has so far hardly taken off – will see the most rapid market growth.

Ever-growing turbines

In just a little over ten years wind power in Denmark developed into a brand new industrial branch that could survive without any subsidies. The size of the largest commercial turbines increased from 20kW in the early 1980s to 2000kW at the start of the 21st century. Five years later turbines twice as big are entering the market – commercial wind turbines are now bigger than the prototypes that were built with huge R&D funds in the 1980s. There is one important difference, however: the commercial turbines of today are competitive and reliable; they work. And for each increase in size, the cost per kWh produced has decreased.

That commercial turbines would grow as big as they are today has not been self-evident. This is due to the fact that when you increase the size of a turbine (height and rotor swept area), costs increase faster than size. If you double the rotor diameter, the swept area will increase fourfold, but the weight will increase eightfold. This is due to the so-called *square–cube law*. If you increase the size of a structure (like a wind turbine) it will increase in three dimensions (the cube), while the swept area only will increase in two dimensions (the square). And since price is proportional to weight, there should be a limit where cost becomes greater than production/income.

Nor is it sufficient to scale up all components to increase the size of a wind turbine – the whole construction and most components have to be redeveloped so that the increase of weight can be minimized and to reduce the loads on the turbine.

BOX 3.1 TYPES OF WIND TURBINES

There are wind turbines of several kinds and for different applications. They can be classified according to size, construction and function. There is no standard for the classification of size, since sizes change so fast; however, in 2006 this could be a reasonable classification:

Micro turbines – very small turbines used to charge batteries on ships and in caravans, cabins, etc. with less than 1kW nominal power and up to 1m rotor diameter.

Farm windmills – wind turbines with 1–15m rotor diameter and 1–50kW nominal power used to produce power for a farm/house without grid connection.

Medium-sized wind turbines – commercial turbines for grid connection, with nominal power of 50–1000kW and 15–55m rotor diameter.

MW turbines – wind turbines for grid connection with 1–2MW nominal power and 55–80m rotor diameter.

Multi-MW turbines – wind turbines with >2MW nominal power, mainly offshore installations.

Wind turbines for water pumping – multi-bladed turbines for mechanical water pumping are common in countries with cattle farming. The power in the wind is converted to mechanical work. They are usually quite small, with 3–5m rotor diameter.

Type	Power	Rotor diameter	Hub height	Use
Micro turbine	<1kW	<1m	–	Ships, caravans, cabins, etc.
Farm windmill	1–50kW	1–15m	5–30m	Off-grid
Medium-sized turbine	50–1000kW	15–55m	30–70m	Grid-connected
MW turbine	1–2MW	55–80m	45–100m	Grid-connected
Multi-MW turbine	>2MW	>80m	>65m	Offshore

This is in fact a great technological challenge, and each time a new generation of wind turbines has appeared on the market, most experts (myself included) have been convinced that this new size of turbine was the optimum size from a cost-efficiency perspective. To make even bigger turbines would be technically possible (many MW turbines had been built as R&D prototypes and some were actually running), but the cost of the produced power would be higher.

However, each time (so far) this has proved wrong – the bigger turbines have in fact turned out to be more cost-efficient. This has been due to two factors: by optimizing components the weight has increased not by the cube, but only to the power of 2.6 (an exponent), and by increasing the hub height more energy has been captured by the turbine, since wind speed increases with height.

Manufacturing strategies

Wind turbine manufacturers have followed two different strategies:

- **Off-the-shelf components** – some manufacturers used a strategy of buying cheap standard components, manufactured in large volumes for general purposes (generators, gearboxes, brakes, etc.), to let subcontractors manufacture components (towers, rotor blades and control equipment) and to just do the mounting of the parts in their own factory. This was the most common strategy up to the middle of the 1990s. In this approach it is the technological know-how that is the main asset of a wind power manufacturing company.
- **In-house production** – other companies adopted the strategy of manufacturing as many components as possible in their own factories so that these specially developed components wouldn't be available to other competing wind turbine manufacturers.

Now these two strategies have converged: the market has grown so fast and to such an extent that subcontractors are able to manufacture components, especially designed for wind turbines, in long series and using mass production. Some manufacturers have grown so strong that they have even bought out their former subcontractors.

The so-called Danish standard concept, a three-bladed upwind rotor with gearbox, has become the dominant model, as it has turned out to be reliable and cost-efficient. There are, however, other concepts as well. The German manufacturer Enercon's concept with a multi-pole direct-drive generator (without gearbox) and variable speed is definitely competitive. In recent years a hybrid of these two concepts has also appeared on the market – turbines with a one-step planet gearbox combined with a multi-pole PM (permanent magnet) generator. This concept can reduce weight when the size increases.

From the late 1990s the competition between manufacturers became more intense. Growth on the world market has been around 30 per cent, a fast growth rate indeed. The manufacturing companies were introduced on the major stock exchanges, and this put pressure on them to show good profits. New larger models were introduced too fast (an obvious mistake in hindsight), without the two to three years' testing that had been the rule in earlier years. This rapid increase of size had a price, for manufacturers as well as for the owners. Technical problems arose, especially with the gearboxes. Manufacturers had to retrofit and/or replace gearboxes in thousands of turbines. Others had problems with rotor blades, and the US manufacturer Kenetech Windpower went bankrupt because its blades broke and the company could not afford to replace them.

This rush for ever-bigger turbines also cost a lot of money. Many manufacturers got into economic problems, and were taken over by companies with more economic strength. The German company Tacke was bought by the American power company Enron, which itself went bankrupt a few years later in one of the largest bankruptcies in the US. In its aftermath General Electric, a giant multinational company from the US, bought Enron Wind.

In Denmark the company NEG Micon bought the smaller manufacturer WindWorld but was later merged with the largest Danish manufacturer Vestas. In 2004 the multinational German industrial giant Siemens bought another major Danish manufacturer, Bonus.

During the same period the projects have also grown larger. From installations with single turbines at farms in the 1980s, groups of three to ten turbines and even larger wind farms in the 1990s, the focus in Europe from 2000 onwards has been on the development of large offshore wind farms, with utilities and oil companies as investors. Today wind power is a major industrial branch in many countries and has become big business.

Figure 3.1 *Middelgrunden offshore wind farm*

The Middelgrunden offshore wind farm is located just outside Copenhagen, in the harbour area. It consists of twenty 2MW turbines. Half of the turbines are owned by the utility in Copenhagen, the other half by a cooperative with 8000 members (households from the Copenhagen area).

Source: Tore Wizelius

Offshore wind power

Denmark was also the pioneer with offshore wind power. This rather small country had made a plan for wind power development in the early 1980s, and according to this plan offshore installations would be built in the early 1990s. The first offshore wind farm was installed at Vindeby, and a few years later a second one at Tunö Knob. Based on these experiences a plan was drawn up to install several huge offshore wind farms so that Denmark could get 50 per cent of its electric power from the wind by 2030. The reasons for the offshore emphasis were that available land for wind farms onshore was limited and that the wind resources at sea were far better than on land. The cost to install wind power offshore is 40–50 per cent greater, but the increase in production will be similar.

During the last few years several offshore wind farms have been built – Horns Rev, with 80 turbines of 2MW each, and the similarly sized Nysted. Several smaller offshore installations have also been built, not only in Denmark, but also in Sweden, the Netherlands and the UK. Several countries in Europe are now planning for large-scale utilization of wind power offshore, with installations of up to 1000MW that can produce as much power as a nuclear power plant. The most recent 5MW turbines that have been developed are mainly aimed at the offshore market. With the ambitious plans for offshore development and growing markets in many other regions of the world, wind power is likely to sustain a growth rate of 20–30 per cent for another decade.

Technical specifications

Height. The height of a wind turbine is indicated in two ways. *Hub height* is the distance from the ground to the hub of the rotor. The hub height is a little higher than the tower. The *total height* is hub height plus half of the rotor diameter.

Power. The generator's maximum power is called its *rated power* (measured in kW). Sometimes two values are indicated, for example 600/120. This means that the turbine has two generators (or a so-called double wound generator), a small one for modest winds and a big one for strong winds. Rated power is reached at the turbine's rated wind speed (which can vary from 12 to 16m/s on different turbine models); when wind speed is lower, the generator's power is lower than the rated power.

Rotor. The size of the rotor is indicated by the *rotor diameter* (measured in metres), i.e. the diameter of the circle the rotor covers when it rotates, and as *swept area* (m^2), the area of the same circle. The area is calculated by $A = \pi(D/2)^2$, where A = area, D = diameter.

BOX 3.2 THE WIND TURBINE

The wind turbines that dominate the market today have a rotor with three blades that rotate at 15–30 revolutions per minute (the rotational speed depends on the rotor diameter – bigger rotors rotate slower). The wind turbine's main components are its foundations, tower, nacelle (generator, gearbox, etc.), rotor, control system and transformer. The rotor size is defined by the rotor diameter and swept area; the height in metres is defined by the hub height and total height respectively. A typical wind turbine structure is illustrated in Figure 3.2.

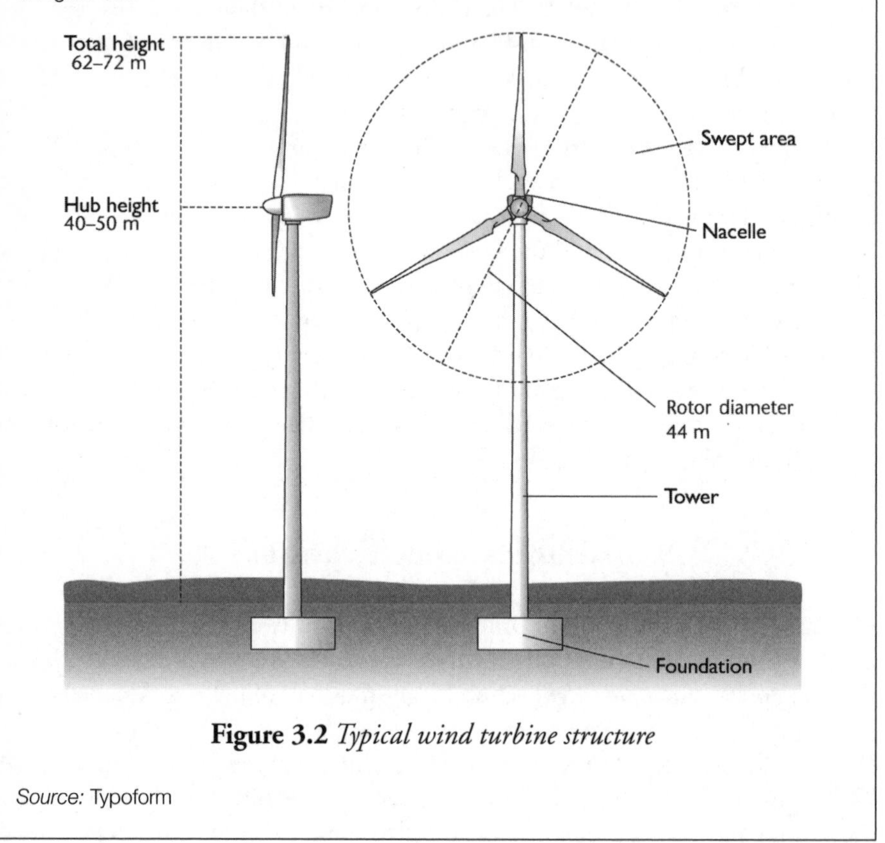

Figure 3.2 *Typical wind turbine structure*

Source: Typoform

Rotational speed. The rotor's rotational speed is indicated by revolutions per minute (rpm). Most older turbines have constant speed; the rotational speed is independent of the wind speed. Wind turbines with two generators use two different constant rotational speeds, a slower one at low winds (for the small generator) and a higher speed with stronger winds (the large generator). Ideally, rotational speed should be proportional to the wind speed. Today most manufacturers therefore use *variable speed*, especially on large models. The rotational speed in this case is indicated by a range, for example 20–35rpm.

Power control. When the wind speed increases above the rated wind speed, the power has to be regulated to avoid damage to the turbine. Two different methods are used for this: stall or pitch or a combination of the two, and active stall.

Wind speed. Three different wind speeds are indicated: *cut-in wind speed*, the wind speed necessary to get the turbine to start and produce power; *rated wind speed*, the wind speed when the turbine reaches its rated power; and *cut-out wind speed*, the wind speed when the turbine's control system will stop the turbine for safety reasons. The cut-out wind speed is usually set to 25m/s.

PART II

Wind Energy

The wind is the renewable energy source that wind turbines utilize. In this part the character and properties of wind are described in Chapter 4, followed by an account of the power in the wind and how its energy content can be calculated. How the power in the wind can be captured and converted into mechanical power is explained in Chapter 6, where the basic aerodynamic properties of blade profiles used by modern wind turbines are also described.

4

The Wind

Wind is air in motion and wind turbines turn the kinetic energy of the moving air into electric power or mechanical work.

To get an idea of how this renewable energy source works there follows a description of how wind is created, the properties of the wind in mid latitudes, and other factors relevant to wind power. In this context it is the wind climate at a specific site or within a region that is of interest (see Box 4.1).

Box 4.1 Weather and climate

Weather is the totality of atmospheric conditions at any particular place and time – the instantaneous state of the atmosphere. The elements of the weather are temperature, atmospheric pressure, wind, humidity, cloudiness, rain, sunshine and visibility.

Climate is the sum total of the weather at a given place during the course of the year and over a period of years. Since the average conditions of the weather elements change from year to year, climate can only be defined in terms of some period of time, some chosen run of years, a particular decade or decades.

Wind climate is the long-term pattern of the wind in a specific site, region or country.

Climate can be studied and analysed at different levels:

- **Macro climate**, large-scale climate patterns on the Earth, continents or parts of continents.
- **Meso climate**, the climate in a country or a region.
- **Local climate**, the climates within a limited area, such as a coastal zone, wood or city block.
- **Micro climate**, the climate at a specific site, a field, a pond or the sunward or shaded side of a tree.

Atmospheric pressure and temperature

Winds are created by differences in atmospheric pressure that in turn are caused by differences in temperature. The atmospheric pressure tends to be equalized by movement of air from regions with high pressure (H) to regions with low pressure (L). Differences in temperature occur because the Earth is round and the angle of the solar radiation that hits the Earth varies. The Earth rotates around its axis, so radiation differs during day and night. The Earth is divided by longitudes that go from pole to pole and latitudes that are parallel to the equator (see Figure 4.1).

The solar radiation will be perpendicular to the equator (at the spring and autumn equinoxes to be exact). If we keep to the northern hemisphere, the angle between the sun and the surface of the Earth will decline the further north you move, so the same amount of radiation will be spread over an ever-larger area. The distance the radiation has to pass through the atmosphere also increases. Therefore the sun will heat the surface much more at the equator than at the Arctic circle.

The rotation of the Earth will have a similar effect. The angle of the sun, and therefore also the solar radiation per unit of area, increases from sunrise until noon and diminishes in the same way from noon to sunset. In the night-time the Earth is

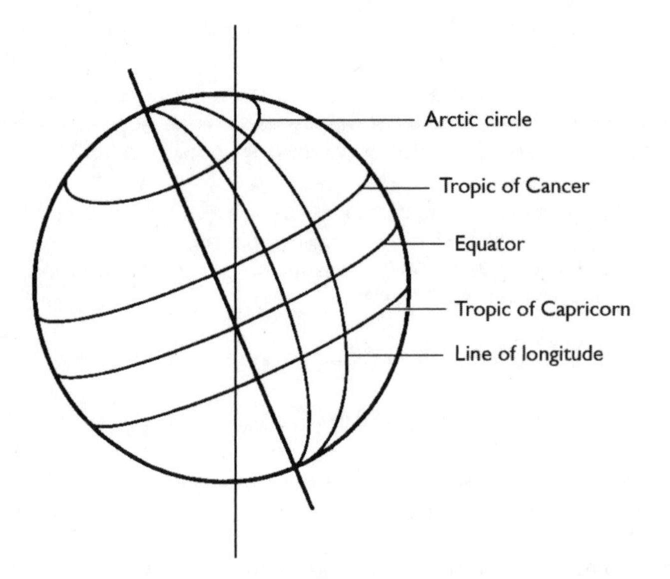

Figure 4.1 *The Earth's coordinate system*

The Earth is divided by longitudes (which circle the Earth with each passing through the North and South Poles) and latitudes (circles parallel to the equator). During a year, when the Earth circles the sun once, the sun moves (seen from a position on Earth) from the southern to the northern tropical circle and back again. At the summer solstice the whole area within the northern Arctic circle has midnight sun, while at winter solstice in the same region the sun never rises above the horizon.

Source: Bogren (1999)

not heated at all; during the night some of the heat that has been stored in the ground and in the sea radiates back into the atmosphere and finally returns out to space.

Finally, the Earth's axis is inclined relative to the plane on which the Earth moves around the sun. This inclination gives us different seasons. At the spring and autumn equinoxes the sun is perpendicular to the equator and day and night have equal duration. From the spring equinox to the northern hemisphere midsummer the sun moves (from a position on Earth) north to the Tropic of Cancer, 23.5 degrees north; it returns to a position over the equator at the autumn equinox and moves on south to the Tropic of Capricorn, 23.5 degrees south, at the winter solstice and then moves back towards the equator (see Figure 4.2).

During a year, when the Earth passes once around the sun, the height of the sun varies and so the temperature in different parts of the Earth will vary. Far north and far south the duration of daylight also varies a great deal.

Several other factors also contribute to increasing or moderating the changes in temperature. Oceans cover a large part of the Earth, and water has quite different properties for storing heat than solid ground. Solar radiation can penetrate at least ten metres down into water, and water can store heat better than soil – it is heated more slowly but it cools off more slowly too. On land the sun only heats the soil to a depth of a few centimetres, and that heat is released again quite quickly during the night. Oceans and lakes therefore have a balancing influence on temperature. Ocean currents as well as winds contribute to the transport of heat between different latitudes and to evening out the differences in temperature. The climate close to oceans, a *maritime* climate, is quite different from the climate in the interior of large continents, a *continental* climate.

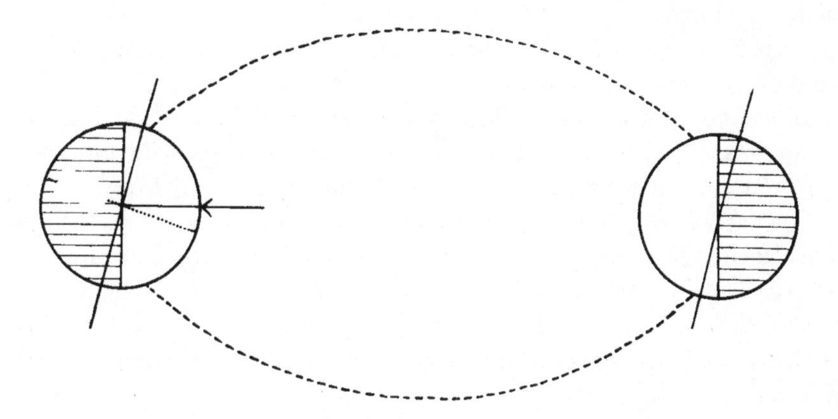

Figure 4.2 *Seasonal changes*

The position of the Earth in relation to the sun in July (Aphelium) is shown on the left; that in January (Perihelium) on the right. Since the axis of the Earth has an inclination, the northern hemisphere will get more sunshine in the summer than in the winter.

Source: Bogren (1999)

Finally there are *clouds*, floating in the air, which consist of water droplets or ice crystals. In the daytime clouds can shelter the Earth's surface from direct sunlight and reduce the heating of the ground. In the night the clouds act as a cover that reflects heat radiation back towards the Earth, keeping it warm. Night temperatures are generally warmer when it is cloudy than when you can see the stars.

Air movements

These ever-changing temperatures give rise to certain more or less regular patterns for the movement of air. At the macro level we have winds that vary with the seasons, like the trade winds and monsoons. In the northern mid latitudes, for example in the UK, northern Europe and Scandinavia, the weather appears to change rapidly and be hard to predict, but the climate in this region is part of a larger pattern characterized by moving low pressure areas, *cyclones*, that are created out in the Atlantic and then move across the British Isles and Scandinavia. These low pressure areas are created where the hot tropical and the cold polar air masses meet.

The winds start when air begins to move from areas with high atmospheric pressures to areas where the pressure is lower. The wind always moves from high pressure (H) to low pressure (L). And these differences in atmospheric pressure are in turn created by differences in temperature.

In coastal areas there are local winds, the sea breezes and the land breezes that give a good illustration of how winds are created (see Figure 4.3).

Weather systems where air circulates between high pressure and low pressure occur in the global weather system (trade winds), regional systems (monsoons, cyclones) and in local systems (sea and land breezes).

If the Earth was flat, didn't rotate and was turned perpendicular to the sun, such pressure differences would be equalized very fast. But since the Earth in fact is round and rotates, the winds don't move along a straight line from high to low pressure areas. The Earth's rotation creates forces that make the wind move toward the low pressure centre in a circle, or a spiral, thus creating a cyclone.

The difference in atmospheric pressure between two areas creates the *pressure gradient force*. On meteorological maps the atmospheric pressures, measured in hectopascals (hPa), are drawn as isobars, lines that connect points with the same air pressure. Around a low pressure zone the isobars form more or less regular circles. When the distances between isobars are small, the pressure gradient is strong (and wind speeds higher).

Gravity (g), which pulls the air towards the Earth's surface, is a better-known force. In the layer closest to the Earth, the wind is also influenced by *friction* against the surface. At a certain height, this influence is negligible; this 'undisturbed wind' is called the *geostrophic wind*. The distance from the ground to the geostrophic wind varies, depending on weather conditions and surface roughness. High friction increases the height of the geostrophic wind (see Box 4.2).

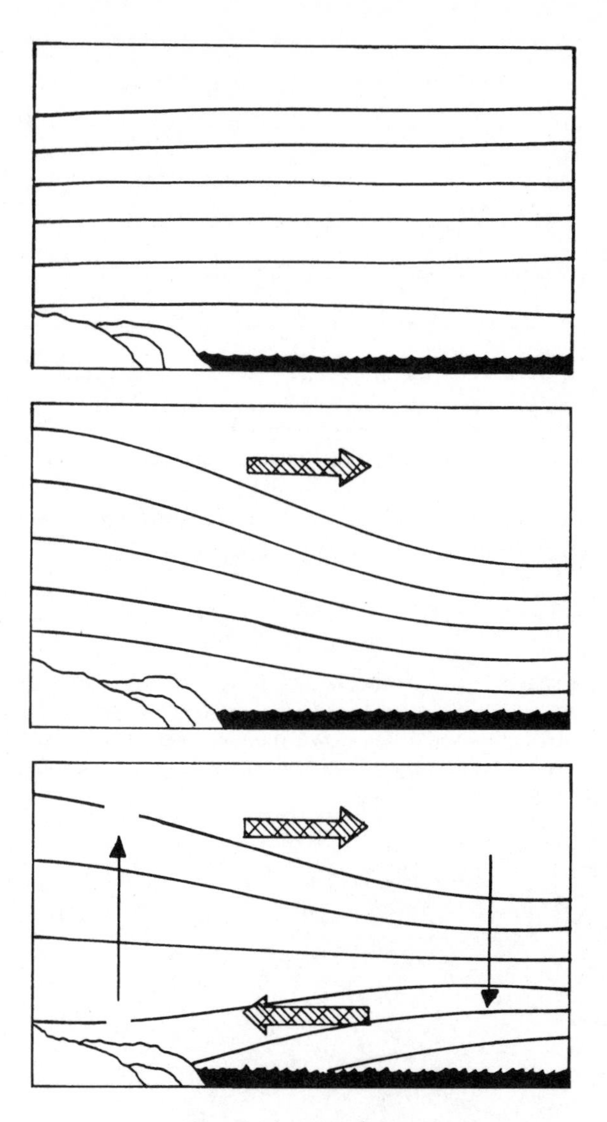

Figure 4.3 *Sea breeze*

When the temperature of the ground and the sea surface are similar, the air pressure above land and sea will be similar as well, and there are no local winds (top diagram). On a sunny day land is heated faster than water. The temperature of the air inland increases, its volume increases, and the air gets less dense and starts to rise upwards in the atmosphere, resulting in a fall in the atmospheric pressure at ground level (middle diagram). The colder air over the sea starts to move towards land to even out this difference in air pressure (the *pressure gradient*). At a certain height above ground level the situation is reversed; the atmospheric pressure is higher above land than over the sea. At that height the air starts to move in the opposite direction, towards the sea (bottom diagram). Thus the air begins to circulate in a local circulation. In the evening the ground loses temperature faster than the sea; the same mechanism then creates a circulation the other way around (the land breeze). These local winds are created by temperature differences between land and sea.

Source: Bogren (1999)

BOX 4.2 GEOSTROPHIC WIND

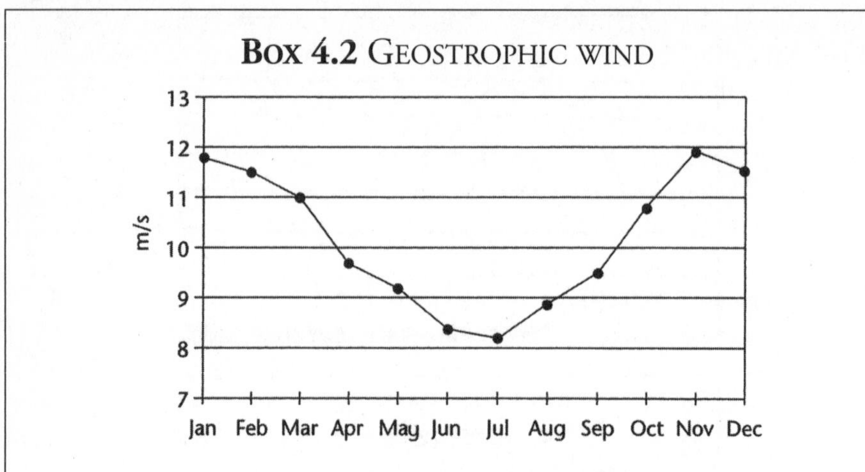

Figure 4.4 *Annual variations of the geostrophic wind south of the island of Öland on the Swedish east coast*

Source: Professor Smedman, Uppsala University

The mean wind speed of the geostrophic wind in Sweden is 10m/s. But just like the wind speed close to the surface, the mean wind speed of the geostrophic wind will vary during different seasons. In winter it is 12m/s (monthly average) and in summer it can go down to 8m/s. Over Sweden the geostrophic wind is strongest in the south and west and decreases towards the north.

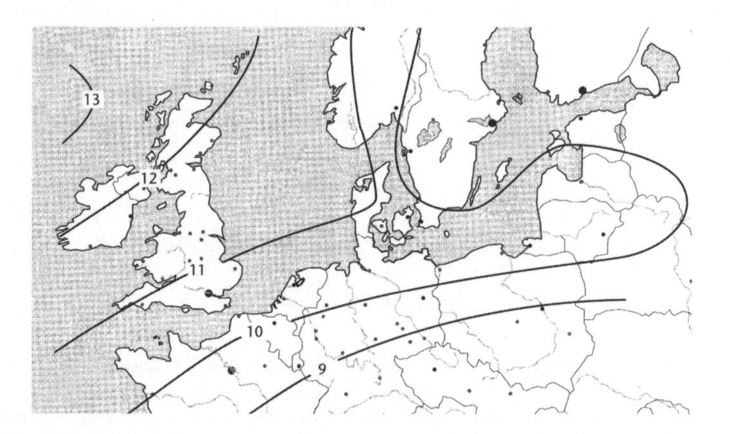

Figure 4.5 *The mean wind speed of the geostrophic wind over northern Europe*

Source: Troen and Petersen (1989)

The geostrophic wind is strongest over the Atlantic and the North Sea and decreases over land. This is not due to friction against the ground, but depends on the large-scale weather systems, where areas with high and low pressure are formed and how they move.

The force that makes the wind bend off from a straight line, and that makes the trade winds blow towards the equator from the north-east and south-east instead of in a perpendicular direction, is called the *Coriolis force*. This force diverts the movement of the air to the right in the northern hemisphere. The Coriolis force is always perpendicular to the direction of movement and proportional to the wind speed. It also varies with latitude. At the equator it is zero and it increases towards the poles. Another better known (but not as strong) force acts in the same direction, this is *centrifugal force*.

A wind that blows from the northern hemisphere perpendicular to the equator is diverted to the right, so the wind will come from the north-east. Winds that blow north from the equator will also be diverted to the right and will thus come from south-west instead of from south. In the mid latitudes there is also a so-called *west wind belt*, and it is the Coriolis force that makes the winds, and the low pressure zones, move eastward from the Atlantic towards Europe. If the earth rotated in the opposite direction, these lows would move towards North America instead.

If we assume that we have a situation where a low pressure zone appears in the north and a high pressure zone in the south, with parallel isobars from west to east, then the pressure gradient force will set the air in motion and the wind will blow from south to north. A so-called *air package* will accelerate northward. The pressure gradient pulls the air perpendicular to the isobars. As wind speed increases, the Coriolis force will divert the moving air to the right (east). The higher the wind speed, the more it will be diverted. Finally a balance will be created where the pressure gradient force and the Coriolis force are equal. And the wind that has this condition, when these forces are in balance, is called the geostrophic wind. This definition corresponds well with the undisturbed (by friction against the Earth's surface) wind, above the boundary layer that was described earlier in this chapter.

Friction against the Earth's surface

At lower heights, the friction against the Earth's surface influences the wind. The friction force acts in the opposite direction to the wind's movement. The friction slows the wind and creates an imbalance between the pressure gradient force and the Coriolis force, so the wind moves across the isobars (in other words it takes a short cut to the low pressure centre). Friction will be stronger closer to the surface. The consequence of this, of course, is that the wind speed will decrease the closer to the surface you get, also that the wind direction will change, more and more across the isobars closer to the surface. This change of wind speed and wind direction is called *wind shear*.

Wind turbines have grown fast, to become higher and higher, but they will always stay within what is called the *friction layer* of the atmosphere. For wind power it is winds up to 150–200m above ground (or sea) that are interesting. At these heights the wind is influenced by local conditions, what the terrain looks like at the actual site, and also by the character of the terrain in an area of approximately 20km radius around the site. The character of the Earth's surface will influence the strength of the friction force; it is, for example, lower over an open field than over a hilly forest.

At the surface, the wind speed is always zero. This does not mean that the molecules of the air don't move, but that the sum of their movements in different directions adds up to zero. The wind speed then increases with height. How rapid the increase with height is depends on the friction against the Earth's surface. On an open plain with low friction, the wind will not be retarded very much, and the increase with height will not be very big. Over a surface with many obstacles, buildings, woods and other structures, the wind will be more retarded, so the wind speed will increase more with height. The relation between wind speed and height is called the *wind profile* or *wind gradient* (see Figure 4.6).

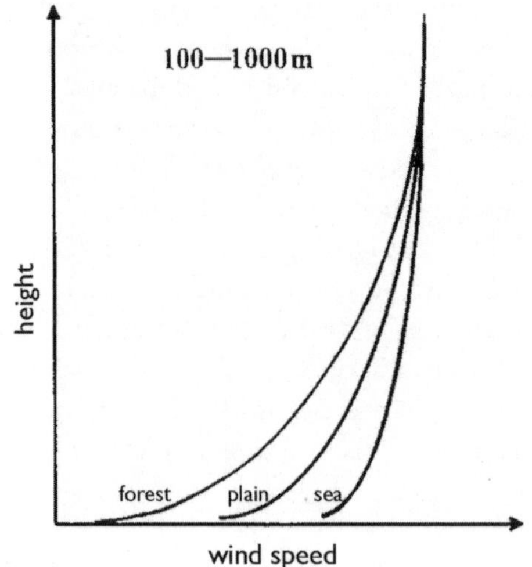

Figure 4.6 *Wind gradient*

The wind gradient, sometimes called wind profile, is a graph that describes the relation between wind speed and height above ground level (agl). The form of this profile depends on the friction of the Earth's surface. Water has low friction, so the wind profile over water will be quite vertical (right graph). Over a plain or areas with agricultural fields, the friction will be stronger and the graph will be more curved (middle graph). Over a forest, the friction will be quite strong and the profile will be more curved than over a plain (left graph).

If you look at the graph from a measuring mast that has anemometers at different heights, in real time the profile does not look as even and smooth as the graphs above. For short periods the wind speed at, for example, 20 metres can be stronger than at 40 metres, and so on. The form of the profile will change continuously. The wind profile actually describes a long-term average of the relation between wind speed and height. At a specific site the wind profile will look different from different wind directions, since it is the character of the terrain in the area that the wind comes from that decides the form of the wind profile.

Source: Tore Wizelius

Turbulence

Since air is completely transparent, it is hard to observe how it moves. However, you can see 'fingerprints' of the air's motions, for example leaves that swirl around on the ground and end up in a heap. A more common sight is to see neatly heaped leaves spread out by the wind so that you have to do the work all over again. It is easier to observe what happens when a stream of water meets an obstacle – around a stone whirls are formed, for example. The air reacts in a similar way: when the wind hits an obstacle, air whirls or waves are formed. The air will then not move parallel to the ground (laminar wind), but in different directions around the prevailing wind direction.

These waves in the air can have a wavelength of several hundred metres. Whirls of air can also be very large, but they will be broken down successively into smaller whirls and finally to movement on the molecular level and turn into heat. When wind is measured, these waves and whirls appear as short variations of wind speed, or *turbulence*.

Differences in temperature in the air can also create turbulence and reduce the wind speed. If the air close to the ground is warmer than at higher levels, and the temperature decreases relatively rapidly with height, warm air will rise upwards. The horizontal wind will then meet air that is moving in a vertical direction, and this creates turbulence. When the wind passes through the rotor of a turbine, a very strong turbulence is created. This whirling wind on the lee side of the rotor is called *wind wake* and influences the wind speed up to a distance of 10 rotor diameters or more behind the turbine.

When the wind blows from the sea and continues over land, the friction against the surface will change at the shoreline. The air that moves close to the surface will be retarded and start to rotate. This turbulent air will move upwards when the wind moves further in over land. An *internal boundary layer* has been created. The wind speed below this boundary layer (the boundary between the laminar wind from the sea and the more turbulent wind over land) will decrease, while the height of the boundary layer that is formed will stabilize. The wind gradient will also change as a consequence of this.

If the wind blows from land out to sea, the wind close to the water's surface starts to accelerate. Every time the character of the landscape (and thus the surface friction) changes, a new internal boundary layer is formed, which will change the wind speed to a level that corresponds to the friction of the underlying terrain. There is, however, a transition zone: it always takes some time and distance before this change of friction is transferred to higher layers, and there will be a delay between the wind energy content and changes in the surface friction (see Figure 4.7).

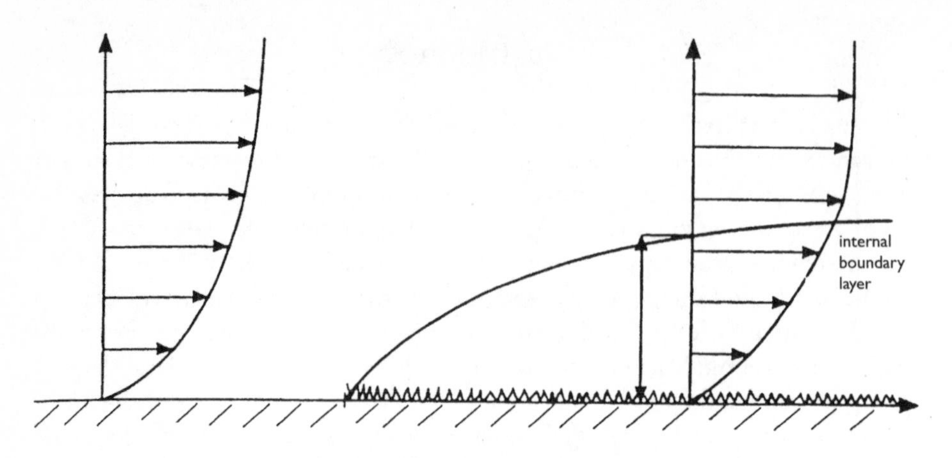

Figure 4.7 *Internal boundary layer*

The friction against the surface affects the wind. When the wind moves from the sea (left) to land (right) the turbulence will increase, so the wind speed close to the ground will decrease. An internal boundary layer is formed between the turbulent wind and the laminar wind from the sea at higher levels. The height of the boundary layer will stabilize after a while and remain constant until the surface roughness next changes. The wind profile also changes: over land it is more curved than over sea. The three arrows closest to the surface, which represent wind speed at these three heights, are shorter over land than at sea. Compare also to the wind profiles in Figure 4.6.

Source: Bogren (1999)

Roughness classes

Terrain is classified into different *roughness classes*. Five different classes are used, from 0 to 4. Open water is 0, open plain 1 and so on to roughness class 4 for large cities and high dense forest (see Table 4.1). Sometimes the expression *roughness length* is used. This length has nothing to do with the length of the grass or the height of buildings; it is a mathematical factor used in the algorithms for calculations of how the terrain influences the wind speed. A description of how a roughness classification is made is given in Part V, Wind Power Project Development.

Hills and obstacles

The wind is also influenced by different kinds of *obstacles*. The impact depends on the height and width of the obstacle, and also by their so-called *porosity* (how much wind can pass through the obstacle). If you want to protect yourself from the wind in your garden, for example, a hedge or a fence that allows some air to pass through is much more efficient than a brick wall or a tight fence. When the wind hits a wall strong turbulence is created behind it. If some of the wind can pass through the fence, there will be much less turbulence. A building or a tree-lined country lane will influence the wind in front, behind and above the obstacle.

Table 4.1 *Roughness classes*

Roughness Class	Character	Terrain	Obstacles	Farms	Buildings	Forest
0	Sea, lakes	Open water	–	–	–	–
1	Open landscape, with sparse vegetation and buildings	Plain to smooth hills	Only low vegetation	0–3/km^2	–	–
2	Countryside with a mix of open areas, vegetation and buildings	Plain to hilly	Small woods, alleys are common	up to 10/km^2	Some villages and small towns	–
3	Small towns or countryside with many farms, woods and obstacles	Plain to hilly	Many woods, vegetation and alleys	Many farms >10/km^2	Many villages, small towns or suburbs	Low forest
4	Large cities or high dense forest	Plain to hilly	–	–	Large cities	High dense forest

According to a simple rule of thumb, an obstacle creates turbulence to double the height of the obstacle, starts at a distance of twice the height in front of it and continues for 20 times the height behind it (see Figure 4.8).

A more exact estimation of how wind speed is influenced at different distances and heights from an obstacle can be made using a diagram from the European Wind Atlas (see Figure 4.9).

Hills also have an impact on the wind speed. A smooth and not too steep hill makes the wind accelerate and the wind speed will increase up to a certain height

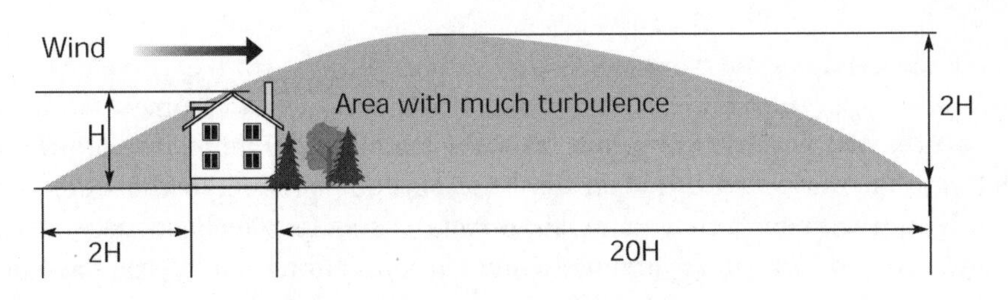

Figure 4.8 *Turbulence from obstacles*

Close to an obstacle, turbulence will increase and wind speed decrease. The turbulence is stronger and is spread further on the leeside of an obstacle, but turbulence will also appear on the side where the wind comes from, since the obstacle interferes with the wind flow. The areas with turbulence will of course vary with the wind direction.

Source: Gipe (1993)

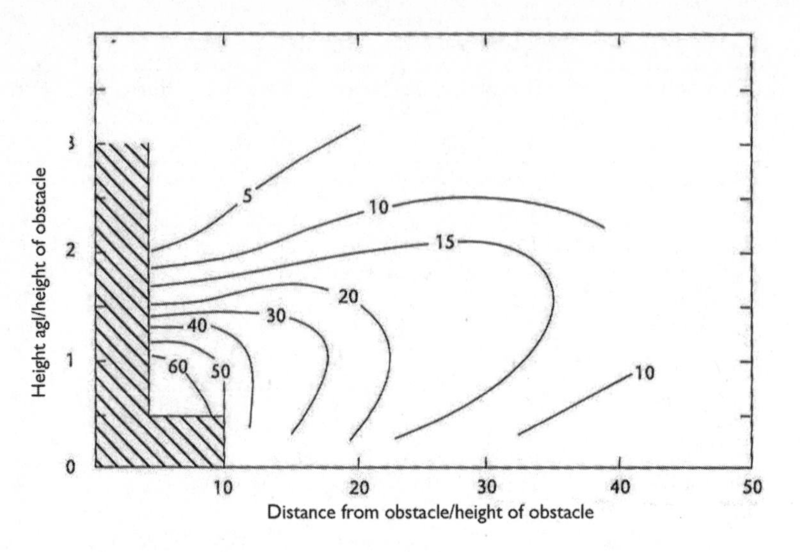

Figure 4.9 *Diagram for estimation of impact from obstacles*

This diagram gives a general picture of the relationship between an obstacle and the reduction of wind speed at different heights and distances from it. The vertical axis gives the relationship between the height above ground level and the height of the obstacle. The horizontal axis gives the relationship between the distance from the obstacle and the height of the obstacle. (Note that the rectangular area on the y-axis is not the obstacle, but the area that is too close to the obstacle to make measurements/calculations.)

The figures and curves in the diagram show the reduction of wind speed, as a percentage, at different points in the diagram. If, for example, we want to find out how much the wind speed will decrease at 20m agl (*H*) 100 metres behind a wall that is 10 metres high, then the ratio of *H* to the obstacle height is 2 (20/10). The ratio of the distance to the obstacles height is 10 (100/10). At the point (10, 2) in the diagram you can see that the wind speed at a height of 20 metres 100 metres behind the 10-metre-high wall will be 10 per cent less than the undisturbed wind speed in front of the obstacle at the same height.

Source: Troen and Petersen (1989)

above the hilltop. On the other side, down slope, the speed will decrease again. If the hill is steep, the air will start to circulate and the airflow becomes turbulent. On a steep mountain slope, turbulence will spread just like at an obstacle and the opposite effect, a decrease in wind speed, occurs (see Figures 4.10 and 4.11).

Terrain with high mountains, deep valleys and steep inclinations is called *complex terrain*. When the wind passes over this kind of terrain, special phenomena occur. This makes it very hard to predict the winds without measurements at the site. Mountain valleys are interesting examples. Along the sides you will get local winds that are created by the differences in temperatures between the top of the hillside and the bottom of the valley, which will vary day and night. In the middle of the valley you may get a so-called tunnel effect – strong winds that follow the direction of the valley, with a maximum wind speed at a relatively low height agl.

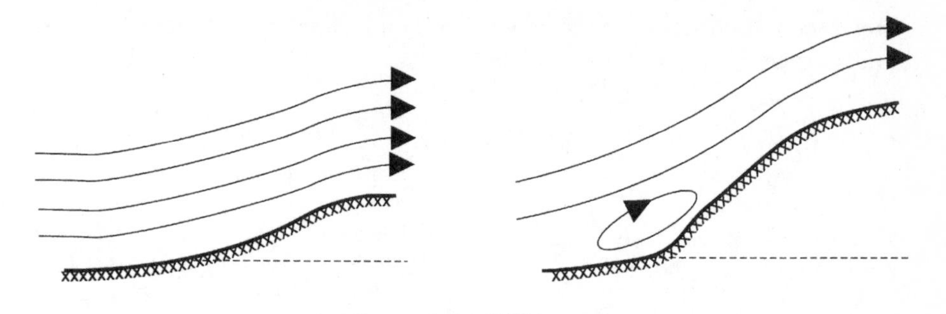

Figure 4.10 *Hill impact*

A smooth hillside that is not too steep makes the wind accelerate and the wind speed and thus the energy content will increase up to a certain height above the hilltop. To get this effect the inclination of the hill should be less than 40 degrees, though if the hillside is uneven and rough the wind flow can be disturbed at inclinations of even 20 degrees. On the leeside of a hill the wind speed will decrease. The wind can, in a similar way, accelerate around the sides of hill, or in a valley, etc.

Source: Bogren (1999)

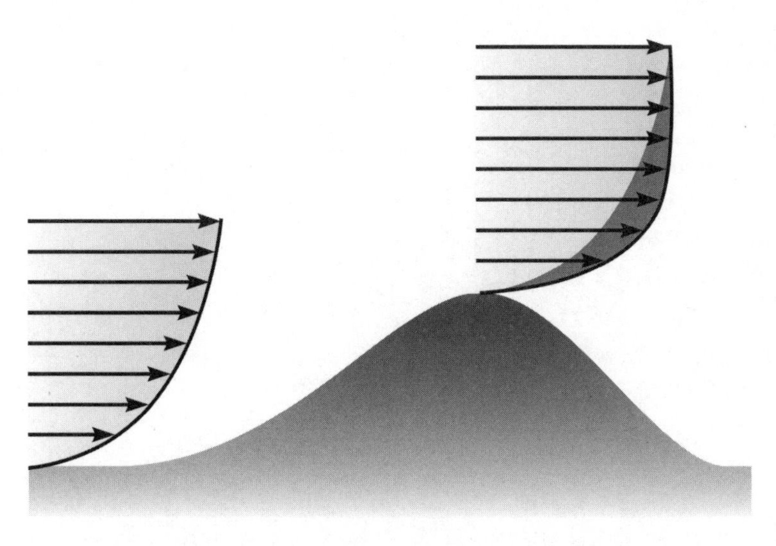

Figure 4.11 *Increase of wind speed from hill impact*

When wind passes over a smooth hill, wind speed increases to the top of the hill. The dark area of the wind profile on the hilltop shows the increase compared to the wind profile of the wind in front of the hill (to the left).

Source: Troen and Petersen (1989)

When you develop wind power you are interested in how much energy the wind will contain during the technical lifetime of a wind turbine, say 20–25 years. To get good production and economy turbines have to be sited in areas where the wind contains a lot of energy. To find the right spot in the terrain, you also need to know something about the wind directions. You have to find out the *wind climate* in the area where you want to build wind turbines. How this is done is described in the next chapter, The Power of Wind.

5

The Power of Wind

Wind is air in motion. Since air has a mass, the wind has kinetic energy (the weight of air is a little more than one kilogram per cubic metre). This power can be turned into electric power, heat or mechanical work by wind turbines.

The energy content of the wind at a specific height at a site is usually specified in kWh per square metre per year: the energy contained in the winds that pass through a vertical area of one square metre during one year. This is easy to calculate – you simply multiply the power per square metre by the number of hours in a year (365 days × 24 hours = 8760 hours) and you get the kWh/m^2/year. Be careful, though, not to mix up the concepts *power* and *energy* (see Box 5.1).

Box 5.1 Power and Energy

Power is energy per time unit and is expressed in watts (or kW, MW, GW, etc.). Power is usually notified by the letter *P*; 1W = 1J/s (joule per second).

Energy is power multiplied by the *time* the power is used. A wind turbine that produces 1000kW power for one hour has produced 1000 kWh. If the wind turbine during a year produces 300kW power on average, it produces 300kW × 8760 hours = 2,628,000kWh per year. *Power is energy per unit of time; energy is power times time*.

The consumption of electricity in a household depends on the size of the household and type of dwelling. For Sweden, for example, the following average values are applied:

- House with electric heating: 20,000kWh/year
- House without electric heating: 5000kWh/year
- Flat in apartment building: 2000kWh/year.

Thus a wind turbine with 1MW nominal power that will produce 2GWh/year can supply electricity to:

- 100 houses with electric heating;
- 400 houses without electric heating; or
- 1000 flats in apartment houses.

Sometimes the same measure of the wind is expressed as a *power density* instead. This gives exactly the same information – it is the mean power density in a year expressed in W/m^2. If you multiply this by 8760 hours and divide by 1000 you get $kWh/m^2/year$.

The wind can be very strong and powerful, strong enough to break branches off trees and rip the roofs from buildings. Storms and hurricanes can create serious natural disasters. This is due to the fact that the power of the wind is proportional to the *cube* of the wind speed. When the wind speed doubles, the power increases eight times (see Box 5.2).

Box 5.2 The power of wind

The power of the wind is calculated in the following way:

$$P_{kin} = \frac{1}{2} \dot{m} v^2$$

where:
P_{kin} = kinetic power (energy/sec) W (J/s)
$\dot{m}$ stands for mass flow; $\dot{m} = \rho A v$
ρ = air density (kg/m^3)
A = area (m^2)
v = speed (m/s).

When $\dot{m}$ is substituted by $\rho A v$ you get:

$$P_{kin} = \frac{1}{2} \rho A v^3$$

The density of air varies with the height above sea level and temperature. The standard values used are usually density at sea level (1 bar) and a temperature of 9°C, giving $1.25 kg/m^3$.

The power of the wind per m^2 is then:

$$P_{kin} = \frac{1}{2}\, 1.25\, v^3 = 0.625\, v^3$$

The power of the wind is proportional to the *cube* of the wind speed:

$$\text{For } v = 5 m/s, \quad P = 0.625 \cdot 5^3 = 0.625 \cdot 125 = 78W$$

$$\text{For } v = 10 m/s, \quad P = 0.625 \cdot 10^3 = 625W$$

$$625 = \mathbf{8} \times 78$$

When the wind speed doubles, the power increases by a factor of 8.

This means that even a small increase in wind speed, for example from 7 to 8m/s, gives a comparatively large increase in power: the power of the wind will increase by 50 per cent! (For $v = 7 m/s$, $P = 0.625 \cdot 7^3 = 214W$; for $v = 8 m/s$, $P = 0.625 \cdot 8^3 = 320W$.)

Therefore it is very important to install wind turbines at sites with the best possible wind resources.

The frequency distribution of the wind

If we assume a site where the wind speed is always exactly 6m/s, the energy content of the wind will be:

$$0.625 \cdot 6^3 \cdot 8760 = 1182\text{kWh/m}^2\text{/year.}$$

In real life, however, the wind speed and wind direction change continuously. Some days are completely calm; on other days there are storms. The wind will also change during the day and night, in different seasons and between years. To be able to calculate the wind energy content (or power density) at a site, you first have to calculate the *average* power density.

If you measure the wind at a site during one year, it is easy to calculate the *mean wind speed* for that site by adding all the measured values for the wind speed and then dividing the sum by the number of observations.

If the mean wind speed at a site is 6m/s, you might draw the conclusion that the energy content should be 1182kWh/m²/year, according to the calculation above. Unfortunately it is not that simple. This is due to the fact that the power is proportional to the *cube* of the wind speed. The cube of the sum of the wind speeds $(v_1+v_2+v_3+\ldots+v_n)^3$ is not the same thing as the sum of the cubes of the wind speeds $(v_1^3+v_2^3+v_3^3+\ldots+v_n^3)$.

If you have a site where the wind speed is 4m/s half of the year and 8m/s the rest of the year, the mean wind speed will be:

$$4/2 + 8/2 = 6\text{m/s}$$

$$6^3 = 216$$

$$0.625 \cdot 6^3 \cdot 8760 = 1182\text{kWh/m}^2\text{/year}$$

But: $$\tfrac{1}{2}(4^3 + 8^3) = \tfrac{1}{2}(64 + 512) = 288$$

$$0.625 \cdot 288 \cdot 8760 = 1576\text{kWh/m}^2\text{/year}$$

To calculate the energy content of the wind at a site it is not sufficient to know the mean wind speed, you also need to know the different wind speeds that occur and their duration, in other words the *frequency distribution* of the wind speeds. The energy content of the wind at two sites with exactly the same mean wind speeds can differ quite a lot (see Figure 5.1).

Data on wind speeds are sorted into a diagram, with wind speed on the x-axis and the duration (in hours or per cent) on the y-axis. To calculate the energy content of the wind during one year, you take cubes of the wind speeds, multiply them by the frequency, add them all up and put the sum into the formula above.

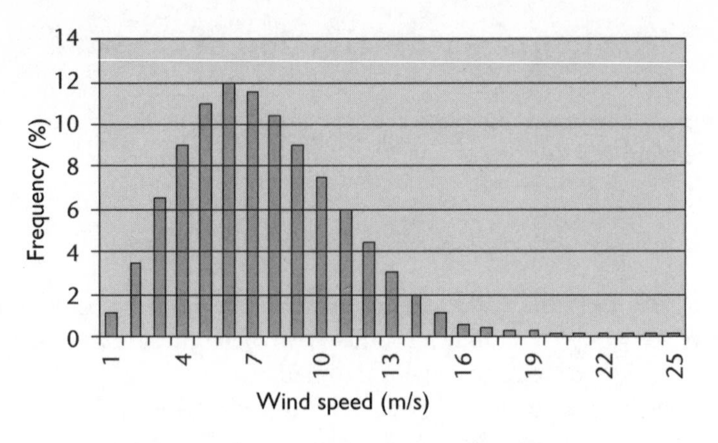

Figure 5.1 *Frequency distribution of wind speed*

A frequency distribution of wind speeds can look like this. The most common wind speeds are in the range 5–8m/s, the wind speed is 6m/s for 12 per cent of the time, and the mean wind speed is 7.5m/s.

Source: Tore Wizelius

The frequency distribution of the wind has proved to fit quite well to a probability distribution called the *Weibull distribution* (see Figure 5.2).

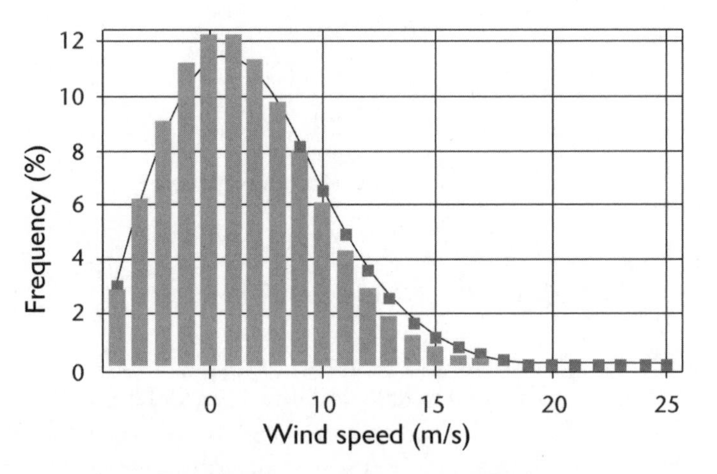

Figure 5.2 *The Weibull distribution*

The frequency distribution of the wind differs at different sites. However, it fits quite well with the so-called Weibull distribution. A Weibull distribution is defined by two parameters: the form factor C and the scale factor A. A Rayleigh distribution is a special case of the Weibull distribution (with form factor a constant C=2).

Source: WindPRO, EMD, Ålborg, Denmark

In the example above 1182kWh/m²/year has to be multiplied by 1.33 to give the correct answer. In practice you can add a factor, the *cube factor*, to the formula $E_{year} = 0.625 \cdot v^3 \cdot 8760$, to calculate the energy content of the wind during a year. The value of the cube factor depends on the frequency distribution of the wind. If the mean wind speed is known but the frequency distribution is not known, a cube factor of 1.91 will in most cases give a good idea of the wind's energy content at a site. This however applies only to places in mid latitudes, the US and most parts of Europe. In areas with trade winds or seasonal winds, or dominant local winds, other values for the cube factor should be used. In Puerto Rico, for example, which has trade winds, the cube factor is 1.4 and in the San Gorgonio pass in California, with local winds, it is 2.4.

Wind measurements

The most correct and detailed information about the energy content (or power density) of the wind at a specific site comes from a measurement mast with an anemometer. The wind directions are also of interest and can be registered by a wind vane. The data are collected in a so-called data logger (see Box 5.3).

With a wind measuring mast wind speed and wind direction at several different heights can be measured. With these data average wind speed over any given period can be calculated; the frequency distribution, power density, energy content and distribution of wind direction can thus be calculated for the measurement period.

A wind turbine has a technical life of 20–25 years. From wind measurements you try to estimate the wind speed and probable frequency distribution for the *coming* 20 years. You want to make a prognosis that is based on solid assumptions.

Even if we measure the wind very accurately for 12 months, what do we actually know? The only thing we know in fact is how the wind has blown during this period. But what conclusions can we make, based on these facts, about the wind's power density in coming years?

Long-term wind climate

Wind speed, frequency distribution and averages vary significantly during different years. And even long-term averages for five- and ten-year periods also vary a lot. How the power density at a site will vary in the long term is important if you plan to use the power of wind, and the longer the period you examine, the less variation you will find, which is very reasonable from a statistical point of view (see Figure 5.3).

BOX 5.3 WIND MEASUREMENTS

Wind measurement equipment usually consists of an anemometer, which measures wind speed, and a wind vane, which measures wind direction. In most countries a national meteorological institute has measured and collected data on the winds since the 19th century, from meteorological stations spread over the country and with local weather observers employed to do this. These have registered wind speed, wind direction, temperature and other kinds of meteorological data several times a day (every four hours, day and night, all year around were registered in Sweden, for example). These data were reported daily to a central institution.

These observations constitute the basis for the *wind statistics* that are used to describe the wind climate in different regions and to create so-called *wind atlas data* that are used to calculate how much wind turbines can be expected to produce at different sites.

Nowadays wind data are registered automatically. Data are continuously stored in a data logger and are then transferred by radio or telephone to the meteorological institute. This is, however, a quite recent development. Not so many years ago, and probably still in some areas of the world, weather observers read the anemometer every four hours, day and night. They observed the anemometer for a couple of minutes and noted the average wind speed for that period.

Both of these methods have their advantages and disadvantages. By manual registration you get less data and there can be systematic differences between different weather observers (in Sweden some 'cold spots' and 'hot spots' turned out to be the creation of weather observers who enjoyed beating 'records'). The advantage is that faults in the equipment can be detected and corrected immediately. With automatic data collection the amount of data will be enormous and the statistical data more exact. Automatic wind measurement equipment registers data continuously and calculates 10-minute averages and one-hour averages. In this way variations in the wind will be described better. There is, however, a risk that the anemometer is giving wrong values, for example if it becomes covered by frost or ice. This can be hard to detect, and you can get very long series of data that are wrong. This means that you can never be sure that wind data actually give a correct description of the wind climate at a site.

The most common type of anemometer consists of three arms with cups on the end that the wind causes to rotate. There are also more modern anemometers that measure the wind speed and direction with light (laser) or sound pulses (sonar). This new technology is, however, still very expensive and not completely reliable.

The power in the wind, energy content or power density can vary between different ten-year periods by as much as 30 per cent (see Figure 5.4).

To get good background data for a prognosis, measured data for a much longer period than one year is necessary. It is, however, not a sensible strategy to measure the wind for five to ten years before a decision to develop a wind farm is taken.

Wind data from a site that has been logged for a shorter period have to be adapted to a so-called *normal wind year*, which is an average for a period of five to

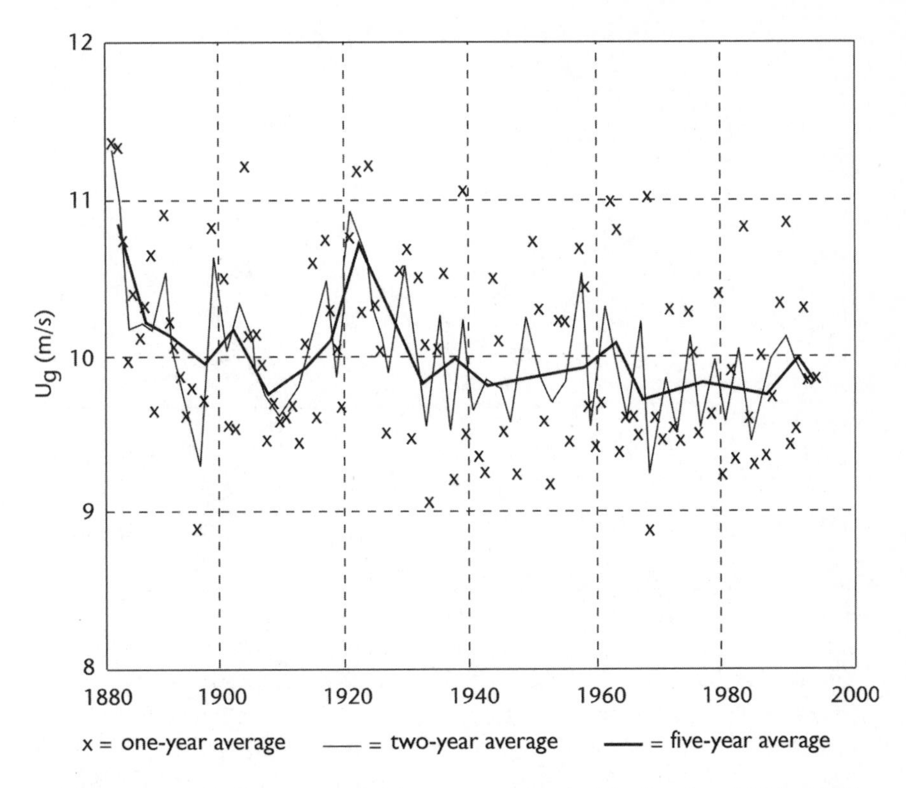

Figure 5.3 *Yearly averages for the geostrophic wind in Sweden, 1881–1995*

This diagram shows averages in Svealand (southern Sweden) from 1881 to 1995, based on observations in Visby, Gothenburg and Lund.

Source: Professor Smedman, Uppsala University

ten years, before it can be used to calculate the wind energy content at the site. The measured wind data have to be compared with corresponding data from the same measurement period in the same region, where long-term data are also available. Then you can check how representative the data from the measurement period are compared to the long-term data from this second measuring mast. Finally you can adjust your own wind data so that they will correspond to a 'normal' year.

The Swedish meteorological institute SMHI has gathered wind data for decades from a large number of meteorological stations in different parts of the country. On most sites it is possible to calculate the energy content of the wind without using your own measuring equipment. Instead you can recalculate the wind data from an existing measuring mast, with long-term data, using the so-called *wind atlas method*. Thus in Sweden it is only necessary to make your own measurements in complex terrain and where available data are unreliable (for example in mountain areas, on large lakes and at sea).

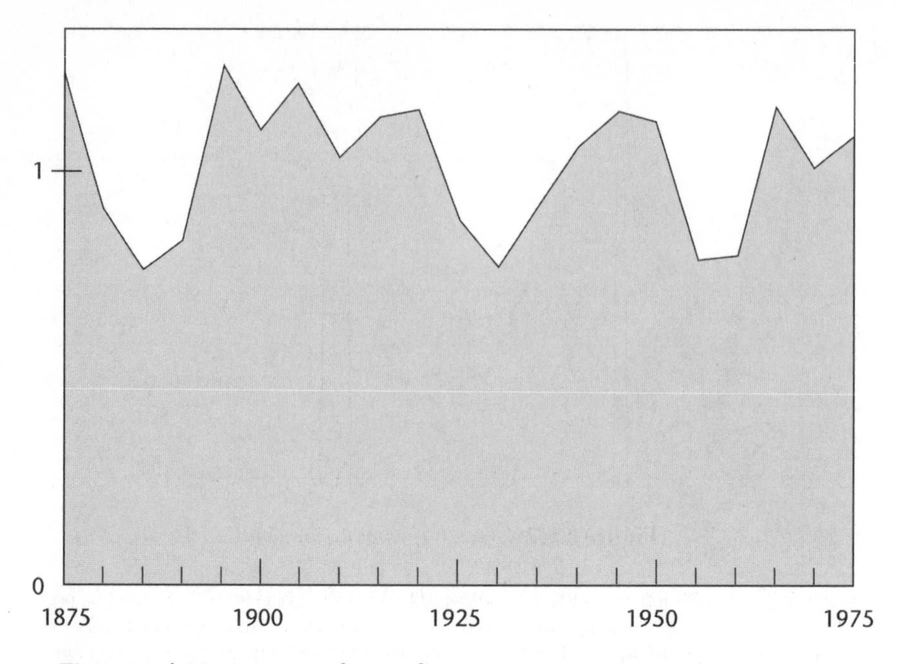

Figure 5.4 *Variations in the wind's energy content, Denmark, 1875–1975*

This diagram shows how the energy content of the wind at Hesselö in Denmark has varied during 5-year periods from 1875 to 1975, compared to the average for the whole 100-year period.

Source: Troen and Petersen (1989)

The wind atlas method

A method to calculate the energy content at different sites was developed by scientists from the Risoe research station in Denmark in the 1980s. The scientists made careful measurements of how the wind was influenced by different kinds of terrain, hills and obstacles. From these empirical data they developed models and algorithms to describe the influence of hills, different kinds of obstacles and topography (see Table 4.1 and Figures 4.6–4.8 in the preceding chapter).

These algorithms were then used in a computer program, WAsP, that can be used to calculate the energy content at a given site using wind data from an existing wind measuring mast with long-term wind data (that are converted to so-called *wind atlas data*) and with information that describes obstacles, hills and the roughness of the terrain within a radius of 20km from the site where the wind turbine will be installed (see Box 5.4).

A wind atlas program works in two steps. The first step is to convert normal long-term wind data (covering five to ten years and both wind speed and direction) from a regular wind measurement mast to so-called wind atlas data. This means that the wind data from the measuring mast are normalized to a common

BOX 5.4 WIND ATLAS PROGRAMS

There are several different computer programs for wind power applications, based on the so-called *wind atlas method*. All of them have a Microsoft Windows interface, are easy to work with and give reliable results. They can be used to calculate how much a wind turbine of a given model can produce at a given site, sound propagation, park efficiency and visual impact. These programs can also be used to make *wind resource maps* and economic calculations.

Wind power researchers at Risoe in Denmark developed the program *WAsP*; EMD (Energi- och miljödata) in Ålborg developed *WindPRO*, which is available in many different languages; and there are two good British programs, *WindFarm* and *WindFarmer*. These powerful programs are very useful planning tools for development of wind power projects. However, ***it is always the quality of the data that are fed into the program that will decide the quality of the output***.

There are also other kinds of programs called *Mesoscale models*. A detailed model of the landscape, information about atmospheric pressure (it is differences in atmospheric pressure that makes the air move) and other data are fed into the program, which then calculates the energy content in the wind for large areas or regions. These are complicated meteorological models and need very powerful computers and a long time to make the calculations.

There are also programs that can make wind and power production prognoses, for a couple of hours to a few days. The program is then connected by modem to a super-computer for meteorological models used by meteorological institutes to make regular weather forecasts.

format, so data from different masts are comparable and can be used by the program.

Wind measurement masts often stand close to buildings and are surrounded by different types of terrain in different directions and often also by hills and mountains. The program can 'delete' the influence from obstacles, orography (height contours) and terrain (roughness), so that the measured wind data are converted to what they would be if the terrain had been plain (roughness class 1) without any hills or obstacles, at 10m agl. The first set of wind atlas data consists of the frequency distribution of the wind in twelve sectors (N, NNW, WNW, etc.) 10m agl in roughness class 1.

These data are then recalculated to other heights: 25, 50, 100 and 200 metres. Together these data describe the *regional wind climate* in an area with a radius of approximately 20–100km (the size of the area depends on local conditions) where the geostrophic winds are the same.

To calculate the energy content of the wind and how much a specific wind turbine can be expected to produce at a given site, the same procedure is followed, but the other way around. Within a reasonable distance from the measuring mast that has been used to process the wind atlas data, the properties of the winds at 200m agl should be the same. By entering data about the roughness of the terrain within a 20km radius of the site, data about hills and obstacles, and finally data about the wind turbine (hub height, rotor swept area, and power curve that describes how much the turbine will produce at different wind speeds), the program calculates the frequency distribution of the wind at the hub height. Finally the program calculates how much the turbine can produce at that site during an average wind year (see Figure 5.5).

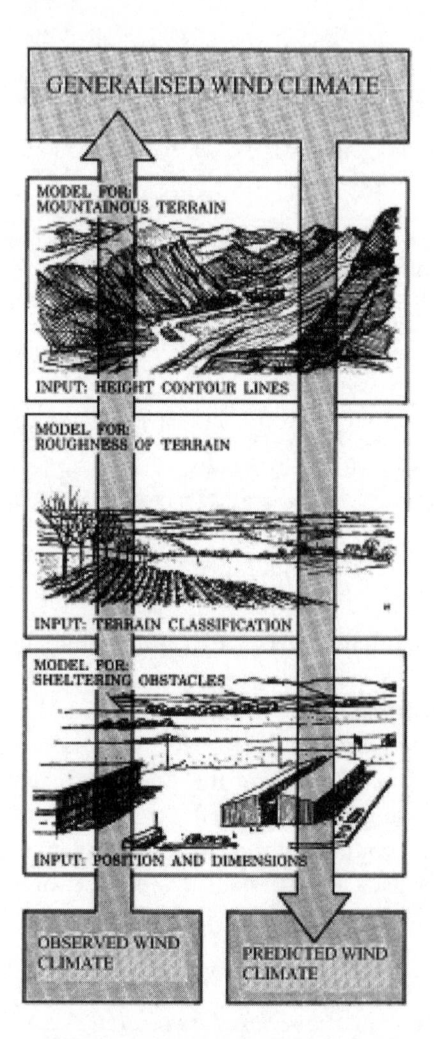

Figure 5.5 *Wind atlas method*

Wind data from meteorological stations are first converted to wind atlas data by clearing away the influence of obstacles, terrain roughness and heights (upwards arrow). These data are then used to calculate the wind climate at specific sites by adding the influence of the specific conditions (obstacles, roughness, heights) at that site (downwards arrow).

Wind resource maps

In many countries meteorological institutes have converted wind data (from five to ten year periods) from a large number of measuring masts to wind atlas data, which can be used as a database in wind atlas programs. Many are already integrated in, for example, WindPRO. Some are available from the Risoe website www.windatlas.dk. At www.rsvp.nrel.gov/wind_resources you can also find wind resource maps for many countries. It is also possible to enter raw data from a wind measuring mast and convert them to wind atlas data. The programs can also be used to produce wind resource maps.

In Sweden the meteorological institute SMHI produced wind resource maps for the southern parts of the country in 1996 and 1997. The energy content of the wind is shown as isolines (kWh/m^2/year) that connect points with the same energy content. Isolines for wind energy are named *isovents*.

Since wind speed increases with height agl, the height for the wind resource map always has to be specified. The standard height for wind measurements is

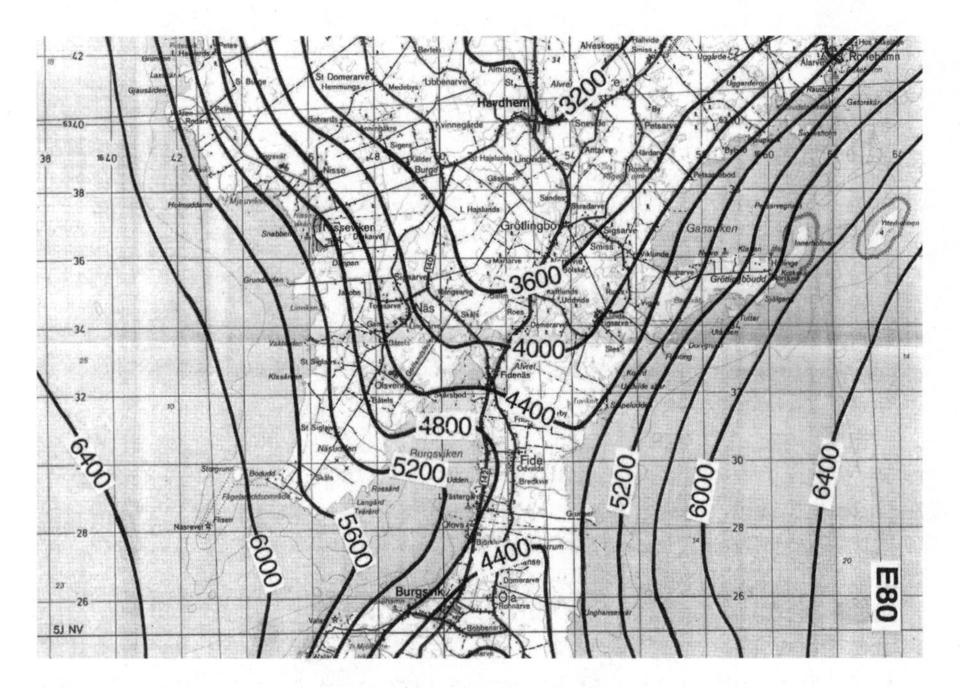

Figure 5.6 *Wind resource map*

On a wind resource map the energy content of the wind is shown by lines. According to this map of southern Gotland, the wind at sea off the west and east coasts, at 80m agl, averages 6400kWh/m^2/year. The energy is reduced as the wind moves in over the island – 4km from the coast (upper middle part of map), the energy has already been reduced by 50 per cent.

Source: Krieg et al (1997)

BOX 5.5 WIND SPEED AT DIFFERENT HEIGHTS

If the average wind speed at a height (h_o) is known and you want to find the wind speed at hub height (h), the following relation can be used:

$$v/v_o = (h/h_o)^{\alpha},$$

where:
v_o is the known wind speed at height h_o and
v is the wind speed at height h.

The value of the exponent α depends on the roughness of the terrain:

roughness class 0 (open water): $\alpha = 0.1$
roughness class 1 (open plain): $\alpha = 0.15$
roughness class 2 (countryside with farms): $\alpha = 0.2$
roughness class 3 (villages and low forest): $\alpha = 0.3$.

Example: if the average wind speed on an open plain (roughness class 1) is 6m/s at 10m height agl, the average wind speed at 50m agl will be:

$$v_{10} = 6\text{m/s}$$

$$h = 50\text{m}$$

$$h_{10} = 10\text{m}$$

$$v_{50} = 6(50/10)^{0.15} = 7.6\text{m/s}.$$

There are different values of α in the literature. These values come from the wind atlas program WindPRO2.4 and the method is called the *power law*. There is another method, but this simple one usually gives the best results when $h \geq 50$m.

10m agl. For wind power the relevant height for calculations is the hub height, which depends on the size of the turbine. SMHI's maps are made for two heights, 50 and 80m agl (see Figure 5.6).

SMHI's wind resource maps have been made with the WAsP wind atlas program from Risoe. The energy content has been calculated for a number of points in a grid covering Gotland. This means that the information is not detailed enough to make calculations for wind turbines at a specific site, but gives a general idea of the areas where the preconditions for wind power development are best.

If wind speed is measured at a site during a shorter period, for example 6–12 months, the wind energy during a *normal* year can be calculated using wind data from a close measuring mast with long-term data available, *if there is a correlation between the wind at the two sites*. The frequency distribution can be assumed to be the same. By calculating the quotient of the measured average wind speed and the corresponding average wind speed from the meteorological station for the same period of time, this quotient can be multiplied by the long-term (five to ten year) average wind speed from the meteorological station. This normalized average wind speed is given the same frequency distribution as the winds at the meteorological station and then the energy content of the wind at the wind turbine site can be calculated.

A more advanced method to do the same kind of long-term correction of the wind at a site is the *long-term Weibull scale method*. The procedure is basically the same, but instead of the mean wind speed the Weibull parameters are adjusted for each of the sectors: $L_C/L_A = M_C/M_A$ where L_C is the long-term form factor, L_A the long-term scale factor, M_C the measured (short-term) form factor and M_A the measured scale factor.

The relation between wind speed and height

As a general rule wind speed will increase with height. How large this increase will be depends on the roughness of the terrain. In areas with high roughness the wind speed will increase more with height than over a smooth terrain. But the wind speed at a specific height, for example 50m agl, will always be higher in an area with low roughness.

With wind turbines you are interested in the wind speed at *hub height*. This height varies for different models and manufacturers, but although available wind data usually represent a different height than the hub height, it is not very difficult to recalculate these for other heights (see Box 5.5).

6
Conversion of Wind Energy

The winds that move above our heads contain a lot of power. To be able to use this as a source of energy, this power has to be 'caught' and transformed to a form that can be used. This can be done with the help of a turbine, which the wind causes to rotate – the turbine turns an axis that can be connected to a millstone, a water pump or an electrical generator.

The wind drives the rotor on a windmill in the following fashion: the rotor blades are inclined in relation to the wind and the moving air pushes against the blades that start to move in one direction with the air moving in the other, an example of action–reaction (see Figure 6.1).

If the rotor is allowed to rotate without any load, it will accelerate, up to a limit. When the speed of rotation increases, the *apparent* wind direction will get closer to the direction of the blade (the chord), until it finally becomes parallel to the blade direction. Then no further power will impinge on the blade, and the speed of rotation will decrease (see Figure 6.2). The apparent wind direction will return to its previous state, the rotor will accelerate again and the procedure will repeat.

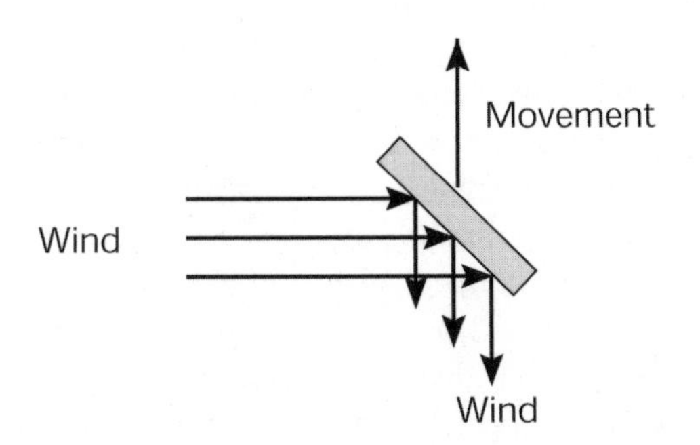

Figure 6.1 *Wind against stationary blade*

When the wind starts to blow on a rotor with simple blades (for example a windmill) while in a stationary state, the air is forced to move in one direction while the blade is pushed in the other and the rotor will start to turn.

Source: Typoform

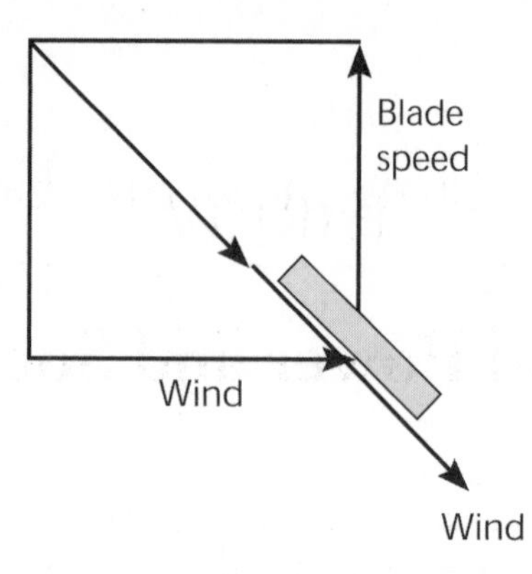

Figure 6.2 *Apparent wind direction*

The blade will react to the resulting wind, which is the sum of the vectors from the horizontal wind speed and the speed of rotation.

Source: Typoform

When the energy in the wind is converted, it is done with a special purpose – to drive a water pump or an electrical generator, for example – which is called the *load*. The axle that is connected to a millstone, a generator or some other device that has to be turned around and offers resistance. It is important to find a good balance between the load/work on one side and the ability of the rotor to catch and convert the energy in the wind on the other. How then should a wind turbine be constructed to be efficient?

The windmill engineers in medieval times were highly skilled in constructing very efficient windmills, but their constructions were based solely on experience. An early example of wind power research was that of Englishman John Smeaton, who in 1759 started to conduct some practical experiments to try to find the most efficient blade angle for a windmill. He built a model of a rotor where the blade angle could be changed and mounted it on a horizontal pole on a device in a barn, so that the pole could be rotated. He attached a wire to the rotor axis that was drawn over a wheel so that different weights could be attached to the other end of the wire. One of his farm hands was then assigned the task of setting the pole with the experimental rotor on the end in motion by pulling a rope that was wound round a vertical axis, so the experimental rotor moved around in a circle inside the barn. The wind speed that hit the rotor would be the same as the peripheral speed of the pole. By changing the angles of the rotor blades and observing which weights the rotor had the power to pull, he managed to find the most efficient blade angle (see Figure 6.3).

The construction of modern wind turbines is based on much more advanced experiments and theories, and there are many questions to tackle: How many

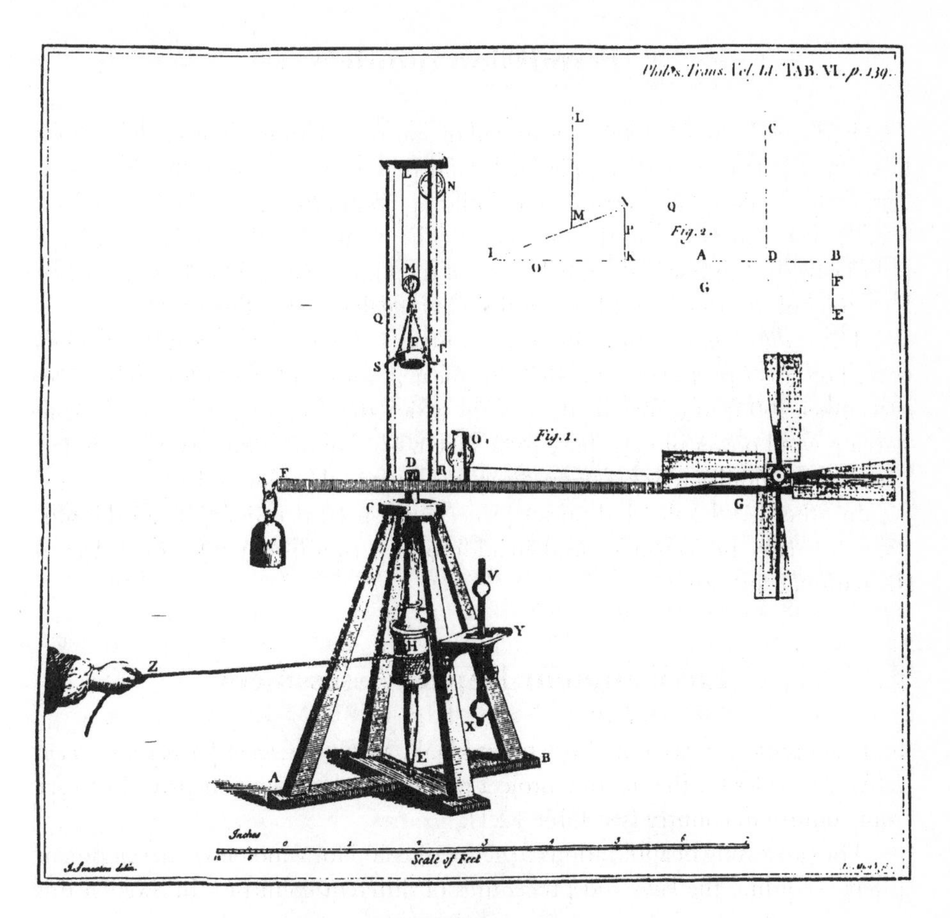

Figure 6.3 *John Smeaton's experiment*

The set-up of Englishman John Smeaton's 1759 experiment to find the most efficient blade angle: the rope pulled around the small rotor at the end of the pole; in the tray (P) weights were placed to measure the power of the rotor. Smeaton tested blades of different forms and angles. He found that the optimal rotor should have an angle of 18 degrees to the plane of rotation in the inner half of the blade, and be twisted to 16, 12 and 7 degrees in the three outer sixths of the blade. He therefore developed a twisted blade.

Source: Hills (1996)

blades should a rotor have? How much of the swept area should they cover? What form should the blades take?

The wind in a stream tube

Wind is air in motion. Air has mass, and the power of wind is the product of the cube of the wind speed and the mass of the air that passes the rotor disc during a specified time. Energy can neither be created nor destroyed; it can only be converted from one form to another; thus to convert the kinetic energy of the wind the moving air has to be slowed down.

To convert all of the wind's kinetic energy, the moving air would have to be retarded completely. That is, however, not feasible. If a rotor slowed down the wind so much that it was stationary behind it, this immovable air would stop the airflow. If you mount a solid rotor it will stop the airflow and the air will escape outside its limits. The wind must be able to pass through the rotor, and also be able to move on behind the rotor. It has to keep some of its wind speed.

The rotor of a wind turbine is a so-called free turbine. The wind and the turbine are unshrouded. In a hydropower station the water is led to a tube and a wall surrounds the turbine so that no water can escape, and thus it will be possible to utilize almost 100 per cent of the kinetic energy of the streaming water. With a free turbine this is theoretically impossible.

Fluid Mechanics is a specialist subject within physics that studies the properties of fluid matter, liquids and gases. *Aerodynamics* is the part of fluid mechanics that is about air. A stream tube is an imaginary tube in the direction of the wind in which the turbine fits (see Box 6.1).

Box 6.1 THE WIND IN A STREAM TUBE

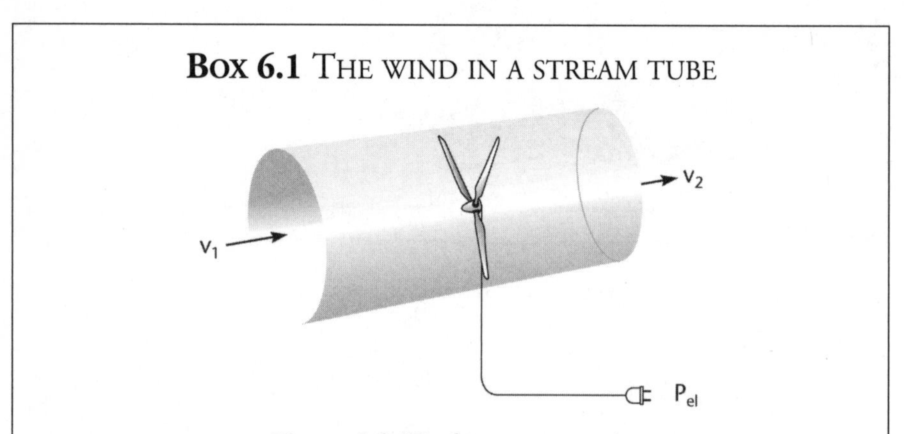

Figure 6.4 *Wind in a stream tube*

Source: Typoform

The power in a stream tube is given by $P = ½ \dot{m} v^2$. If a wind turbine is placed in the stream tube a part of this power will be converted to electric power. The power that enters the tube equals the power that leaves the tube plus the power that has been extracted by the turbine:

$$P_{before} = P_{el} + P_{after}$$

$$½ \dot{m}v_1^2 = P_{el} + ½ \dot{m}v_2^2$$

$$P_{el} = ½ \dot{m}(v_1^2 - v_2^2).$$

The power that can be extracted by the turbine depends on how much the wind speed is retarded. P_{el} can be maximized by choosing a suitable value for the retardation of the undisturbed wind speed v_1.

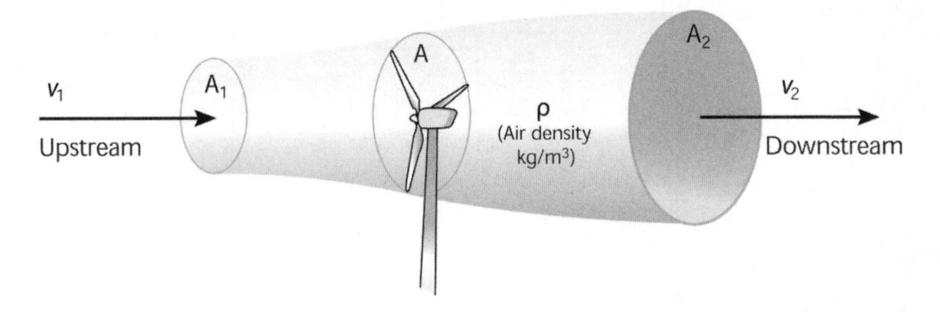

Figure 6.5 *Retardation of wind speed*

The same amount (kg per second) will pass areas A_1, A and A_2 (otherwise air would accumulate in the tube). The mass flow ($\dot{m} = Av\rho$) is the same. This means that $A_1 v_1 \rho = Av\rho = A_2 v_2 \rho$. Since wind speed is retarded ($v_1 > v > v_2$), $A_1 < A < A_2$. The stream tube expands when the wind speed is retarded.

Source: Typoform/Claesson (1989)

How efficiently the power in the wind is utilized depends on how much the wind is retarded by the rotor. If it is retarded too much, or too little, efficiency will be low. The cross-section area of the stream tube in front of the turbine, and thus the mass flow that will pass the turbine, will decrease when the wind speed is retarded (see Figures 6.5 and 6.6).

The modern theory for wind turbines was created by the German scientist Albert Betz from Göttingen and was further developed by Hans Glauert and G Schmitz. Betz showed that a wind turbine is most efficient when the wind speed is retarded by one third just in front of the rotor, and by another third behind the rotor. The undisturbed wind v is retarded by the rotor to $\frac{2}{3}v$ and will decrease to $\frac{1}{3}v$ behind the rotor before it regains its original wind speed due to the influence of the surrounding winds. The power in the wind is used most efficiently in this case – 16/27 (59 per cent) of the power in the wind can be extracted, ignoring aerodynamic and mechanical losses. The rotor of a wind turbine can therefore at most utilize 59 per cent of the energy content of the wind, according to basic theory (see Figure 6.7).

The share of the power in the wind that can be utilized by the rotor is called the *power coefficient*, C_p. Its maximum value is $C_{pmax} = 16/27$ (≈ 0.593). For real turbines C_p is lower, due to aerodynamic and mechanical losses, and the value also varies for different wind speeds. The power that a wind turbine can attain can be expressed as $P = \frac{1}{2}\rho A v^3 C_p$.

How should a wind turbine rotor be constructed? Most old windmills had four rectangular blades that covered about 20 per cent of the swept area. This was mainly due to practical considerations – they were easy to build and worked well. Windmills for water pumping that were developed in the US in the 19th century had blades that covered almost the whole swept area, and the wind passed through

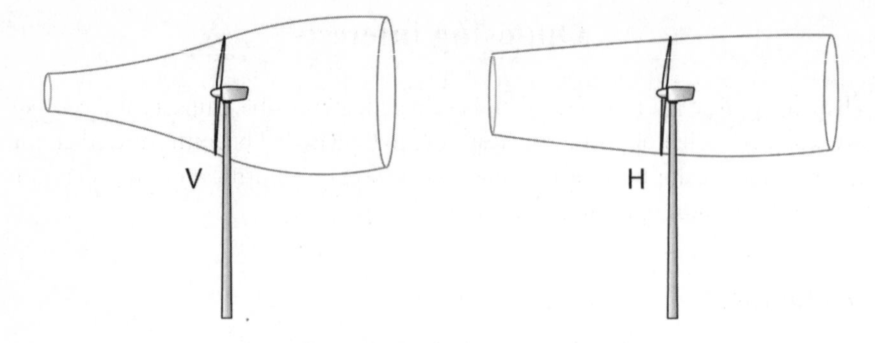

Figure 6.6 *Optimal retardation*

In the example to the left (V) retardation is very strong. The share (%) of extracted power is large, but it is taken from a very narrow stream tube, so the total extracted power will be small. In the example to the right (H) a small share is extracted but from a very wide and thus powerful stream tube. In this case too the total extracted power will be small. The optimal retardation is somewhere between these extreme cases.

Source: Typoform

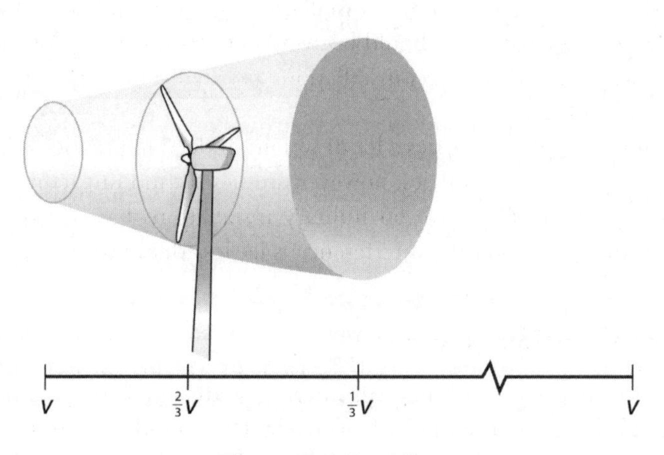

Figure 6.7 *Betz's law*

Alfred Betz showed that half of the retardation happens at the turbine and the rest behind the turbine after the wind has passed through it. The change in wind speed does not happen stepwise but continuously. The undisturbed wind speed v is reduced to $v(1-a)$ when it passes through the turbine and to $v(1-2a)$ some distance behind it (the retardation starts about one rotor diameter in front of the turbine and reaches its maximum about one diameter behind it). The unit a is called the interference factor. If this factor is 0.5, the wind speed behind the rotor will be reduced to zero. This means that $0<a<0.5$. The theoretical maximum share of the power in the undisturbed wind that can be utilized is 16/27 or 59.3%. This maximum is reached for $a = \frac{1}{3}$, which means that the turbine retards the wind speed by $\frac{1}{3}$ at the turbine and by another $\frac{1}{3}$ behind the turbine.

Source: Typoform/Claesson (1989)

slots between the blades. Modern wind turbines use three slender blades that cover not more than 3–4 per cent of the swept area. They are, however, much more efficient than their predecessors.

The optimal rotor has, theoretically, an infinite number of infinitely narrow blades, and obviously you can't manufacture such a turbine. With blade element theory it is possible to calculate the optimum total blade width that can then be divided by the desired number of blades. The fewer blades that are used, the higher rotational speed is needed to get the same efficiency.

The rotational speed of the rotor in relation to the undisturbed wind speed plays a crucial role for the efficiency of the turbine. This is called the *tip speed ratio*. It is a measure of the relation between the tip speed (the speed of the tip of the rotor blades) and the undisturbed wind speed (before it has been retarded by the rotor). From practical experience and by theoretical calculations the optimal tip speed ratio can be calculated for different types of rotors (see Figure 6.8).

The diagram shows that a windmill has a narrow range of tip speed ratios. It is most efficient with a ratio of 2, and the same applies for other older types of wind turbine. Modern turbines with two or three blades have a maximum at a ratio of 10 and 7 respectively. They have a broad span and are quite efficient over a wide range of tip speed ratios.

If the ratio is 1 the blade tip has the same speed as the wind. The apparent wind direction will be 45 degrees (to the plane of rotation), but since the wind is retarded by a third before it reaches the rotor, the apparent wind direction will be 34 degrees. To be able to utilize the power of the wind, the angle of the blades has to be smaller than that, and half as big gives 17 degrees. Windmills, with a tip speed ratio around 2, often have a blade angle (to the plane of rotation) of 15 degrees.

The blade tip of a modern wind turbine has a speed ten times as fast as the wind speed. It may seem implausible for a wind of 5m/s to drive a blade with a speed of 50m/s. This is, however, what actually happens and the following section explains how this is possible.

Aerodynamic lift

If we look at other devices that are driven by the wind, like sailing ships or ice yachts (where the phenomena is more obvious due to lower friction), it will be easier to understand how it works. When a yacht sails in the wind direction, with the wind coming from the rear, it can't move faster than the wind. That is quite obvious. The wind pushes the yacht from behind.

If the yacht beats to windward, that is it sails in a direction where the wind comes diagonally from the front, the wind can't push the yacht. Instead the yacht is pulled by the difference in pressure that is created when the wind passes the sail

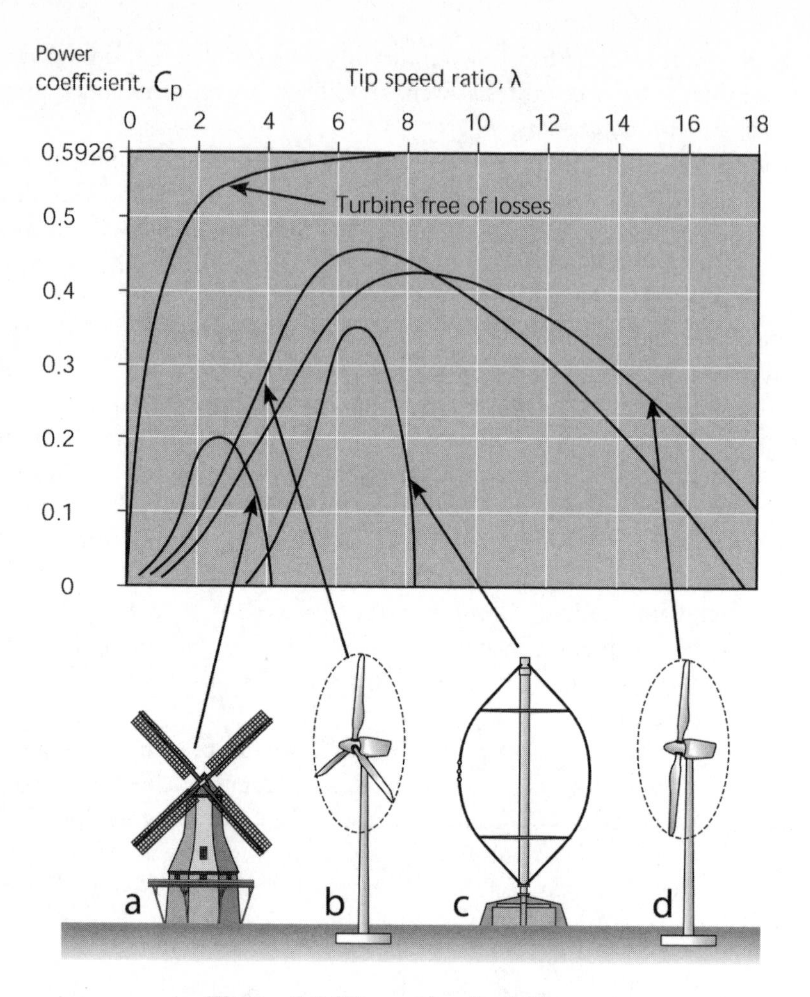

Figure 6.8 *Tip speed ratio diagrams*

The tip speed ratio is the relation between the tip speed v_{tip} and the undisturbed wind speed v_o and is signified by λ (lambda): $\lambda = v_{tip}/v_o$. The power coefficient C_p gives a measure of how large a share of the wind's power a turbine can utilize. The theoretical maximum value of C_p is $16/27 \approx 0.5926$. The diagram shows the relation between tip speed ratio and power coefficient for different types of wind turbines: a) windmill; b) modern turbine with three blades; c) vertical axis Darrieus turbine; d) modern turbine with two blades.

Source: Typoform/Södergård (1990)

(that has a shape similar to an aircraft wing). With a sailing yacht it is not very obvious that it will move faster than the wind, since water creates strong friction. An ice yacht, however, can easily reach speeds of 100km/hour with a wind of 8m/s (30km/hour). The yacht is moving three times faster than the wind, and it is being moved by *lift*, just like an aeroplane is moved upwards and kept in the air by aerodynamic lift.

If you stretch your arm out through the window of a car that is moving at a good speed, you can feel your arm pushed backwards. If you hold the arm straight with your hand parallel to the road, and change the angle slightly, you can suddenly feel that it is drawn upwards. The hand and arm work like the wing of an aeroplane, and with the right angle (of attack) you can feel a strong *lift force*.

These two forces determine the properties of an aerodynamic blade profile (or aerofoil). One force pulls backwards, *drag* (***D***), and another force pulls upward, *lift* (***L***). It is this lift force that makes aeroplanes fly and the rotor of modern wind turbines rotate. (The lift force is actually applied also to so-called drag devices, like windmills and Savonius rotors, but to a lesser degree, otherwise the tip speed ratio would not pass 1.)

The properties of an aeroplane wing are defined by its aerofoil. It is the form of the aerofoil, in combination with the angle of attack, that determines lift and other properties. These properties can be detected by tests in a wind tunnel (see Figure 6.9).

When a stream of air hits the *leading edge* of an aerofoil, some of the air passes over it and some below. When the air stream passes the aerofoil a lift force is created, which depends on the angle of the wind direction in relation to the chord, the angle of attack (see Figure 6.10).

If the angle of attack is too large, the stream of air that passes on the top side of the aerofoil cannot attach to the profile all the way to the back edge (called the *trailing edge*). The flow will stall (whirls are created). When that happens, the lift force will decrease. When the wind speed increases too much this property is used by stall-regulated turbines to limit the power of the rotor (see Figure 6.11).

The strength of the lift force also depends on the form and width of the aerofoil, as well as the wind speed. Modern wind turbines utilize these aerodynamic properties to optimize the aerofoils that are used on the rotor blades of turbines.

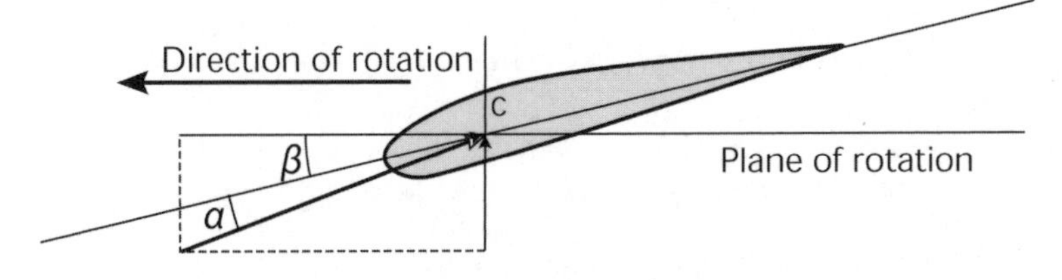

Figure 6.9 *Aerofoil*

An aerofoil is divided by a centreline, the *chord*. The *angle of attack* (signified by α) is defined as the angle between the chord line and the direction of the apparent wind. To get a suitable angle of attack the blades of a wind turbine rotor are set at an angle to the plane of rotation, known as the *blade angle* (or sometimes *pitch angle*). This blade angle is signified by β. The aerofoil also has a centre point, c, which is situated on the chord about 25 per cent from the leading edge of the aerofoil (the distance depends on the form of aerofoil). The sums of lift (and other forces) are summarized (and applied) to this point.

Source: Typoform

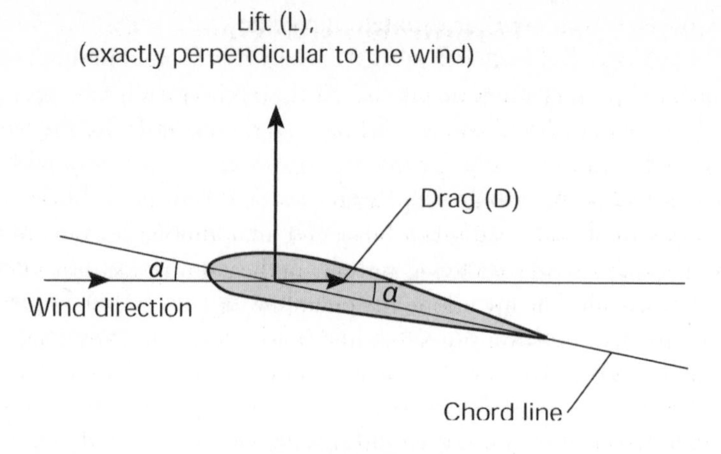

Figure 6.10 *Lift*

Lift ***L*** and drag ***D*** are functions of the angle of attack α. The force ***L*** is always perpendicular to the direction of the apparent wind, while the force ***D*** is applied in the direction of the apparent wind (and therefore perpendicular to ***L***).

Source: Typoform/Montgomerie (1999)

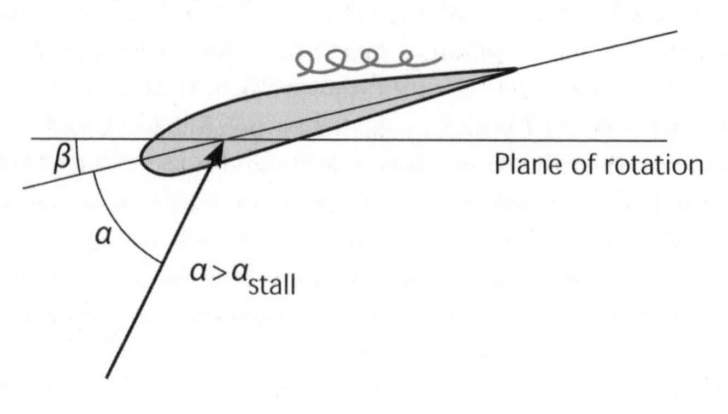

Figure 6.11 *Stall*

When the angle of attack α becomes too large, the airflow cannot follow the topside of the aerofoil all the way to the trailing edge. Eddies are created that reduce lift and increase drag. For a wind turbine with constant rotational speed and fixed blade angles, the angle of attack will increase when the wind speed increases as the angle of the apparent wind direction will increase. Therefore stall can be utilized to reduce lift (and thereby the power) when the wind speed becomes larger than a specified speed.

Source: Typoform/Montgomerie (1999)

PART III

Technology

How wind turbines are designed, the function of different components and technical solutions used by different manufacturers are described in this part of the book. It starts with a brief description of types of wind turbine in Chapter 7, followed by chapters on the wind turbine rotor (Chapter 8), nacelle, tower and foundation (Chapter 9) and electrical and control systems (Chapter 10). How all of these components work together follows in Chapter 11, Efficiency and Performance. The focus is on large commercial wind turbines.

7
Types of Wind Turbine

There are several different design concepts for wind turbines. One basic classification is Horizontal Axis Wind Turbines (HAWT) versus Vertical Axis Wind Turbines (VAWT). Horizontal Axis Wind Turbines can have the rotor *upwind*, that is facing the wind or *downwind* so that the wind will pass the tower and nacelle before it hits the rotor (see Figure 7.1). Today most turbines have an upwind rotor, but there are turbines, from prototypes in the MW-class to smaller turbines with a nominal power of 20–150kW, as well as water pumping wind wheels from the 19th century, with downwind rotors.

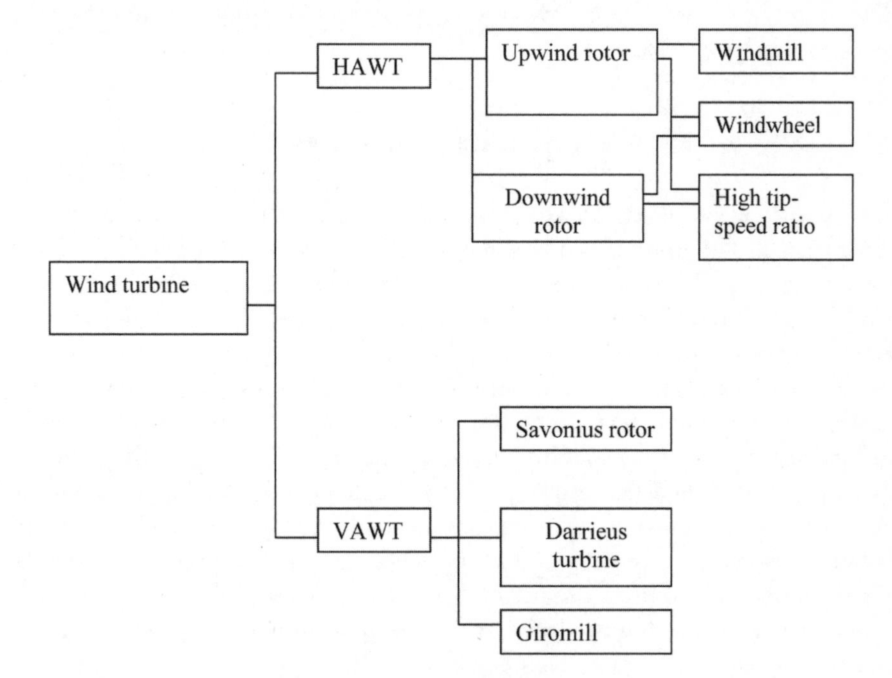

Figure 7.1 *Types of wind turbines*

Source: Tore Wizelius

Horizontal axis turbines

All these types of wind turbines have been built and used in practice. The windmill and the wind wheel, described in Chapter 2, have a long history. The windmill has played its part, but there are around one million wind wheels used for water pumping in use in different parts of the world. They have a very robust design, with quite simple components that are easy to maintain and repair. The advantage with a wind wheel compared to a turbine with few slender blades and high rotational speed is that it starts more easily, since the blades cover a much larger share of the swept area. This is an advantage for a water pump, since it takes a lot of power to get it running.

Turbines with a high tip speed ratio were first used as battery chargers (see Chapter 2). Today they are used to produce electric power that is fed into the power grid, though there are still small micro turbines (with two to six blades) for battery charging. In the 1980s there were many different designs of grid-connected wind turbines, with two or three rotor blades, some with the rotor downwind and others with the rotor upwind. The advantage with a downwind rotor was considered to be that it would automatically adjust itself to the wind direction. However, with sudden changes in wind direction this did not work. At the beginning of the 21st century turbines with a three-bladed upwind rotor completely dominate the market (see Figure 7.2).

Vertical axis turbines

The advantage with a vertical axis wind turbine is that the generator and gearbox can be installed at ground level, making them easy to service and repair. Both the Savonius and the Darrieus turbines are manufactured commercially, but in small models that are used for different niche applications, like battery charging in areas without a power grid (see Figures 7.3–7.7).

The Finnish engineer and inventor Georg Savonius developed a vertical axis wind turbine in 1924, now called the *Savonius rotor*. It consists of a vertical S-shaped surface that rotates around a central axis. By slipping the two halves so that they overlap, and the wind can slip through the middle, efficiency can be increased. Nowadays Savonius rotors are most commonly seen as advertisement posters in front of restaurants and shops. They revolve, but produce no power. There are, however, power producing Savonius rotors as well. They are easy to maintain and reliable, but need a lot of material in relation to the power produced and are not very efficient (see Figure 7.4).

The Savonius rotor is used as a battery charger on lighthouses and telecom masts. It can also be used as a starter motor for a Darrieus turbine (see Figure 7.5).

The French engineer George Darrieus invented this 'egg-beater' shaped wind turbine in 1925. It can have two to four blades, which form bows from the top of

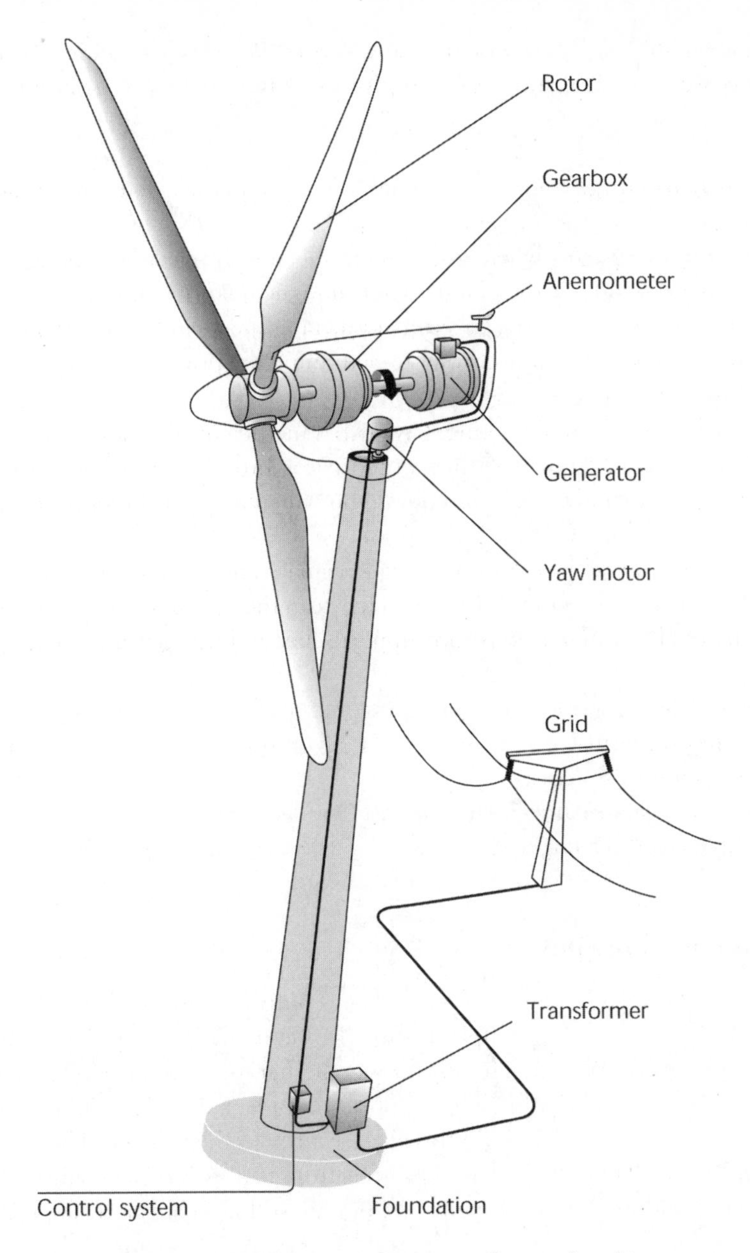

Figure 7.2 *The main components of a wind turbine*

A wind turbine consists of the following main components: foundation, tower, nacelle (generator, gearbox, yaw motor, etc.), rotor, control system and transformer. The turbines that dominate the market today have high tip speed ratios (where the tip speed is 5–7 times faster than the wind), a rotor with three blades and a rotational speed of 10–30 revolutions per minute. Most manufacturers offer several models, with different hub heights and/or rotor diameters, so the turbines can be tailor-made for specific sites.

Source: Typoform (after Tore Wizelius)

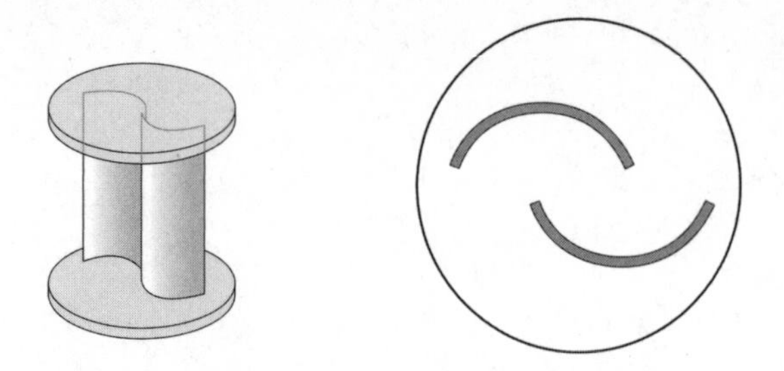

Figure 7.3 *The Savonius rotor*

Source: Typoform/Gipe (1993)

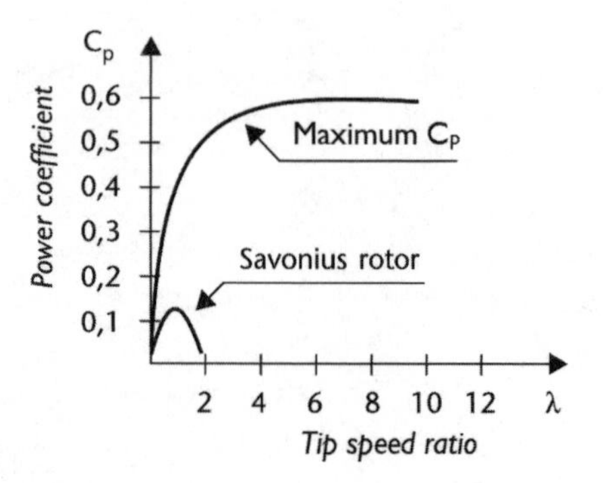

Figure 7.4 *Efficiency of the Savonius rotor*

Source: Claesson (1989)

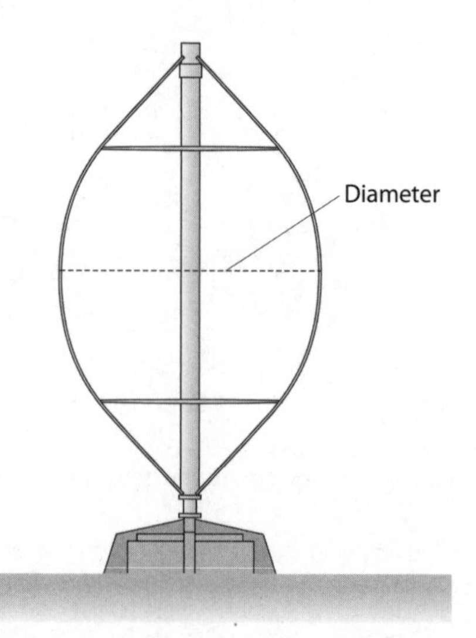

Figure 7.5 *The Darrieus turbine*

Source: Typoform/Gipe (1993)

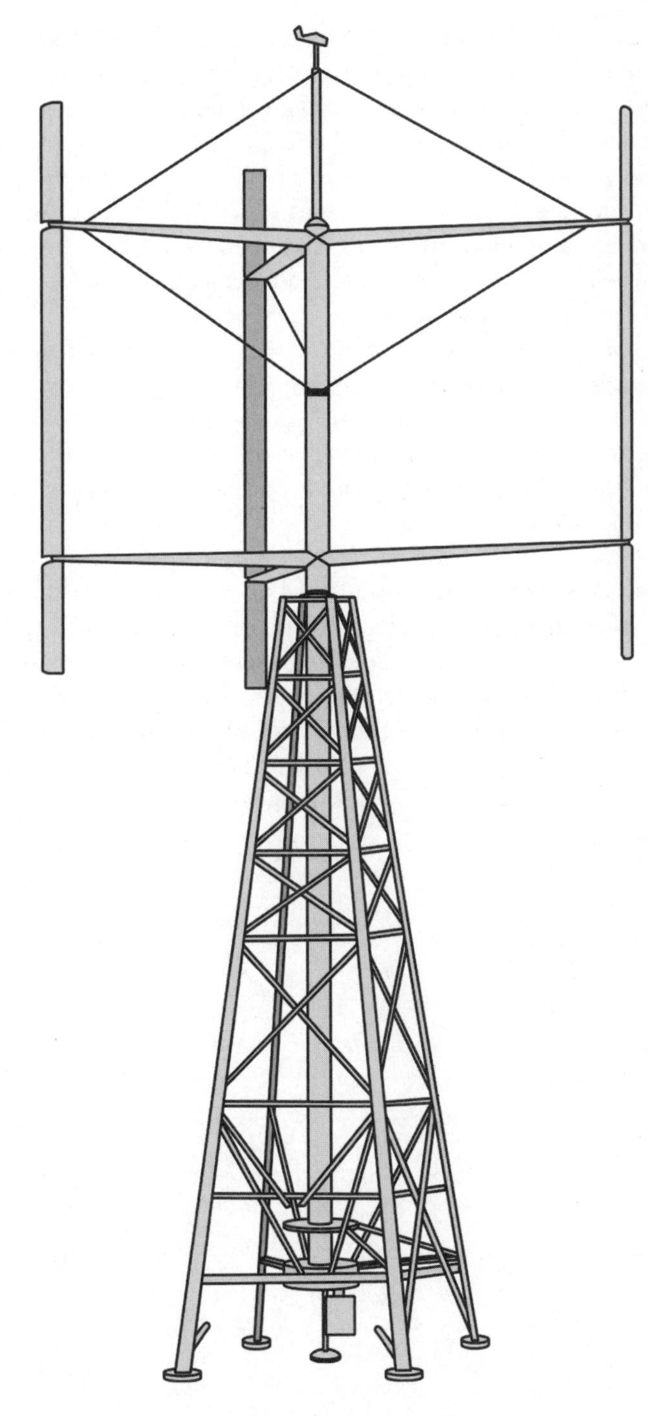

Figure 7.6 *Giromill*

Source: Typoform/Gipe (1993)

the tower to the machinery that is sited at ground level. The blades are symmetrical and very thin. The form directs the centrifugal force to the points where they are connected to the central axis, so that the bending moments are minimized. In most materials the tensile strength is stronger than the bending strength. A *Darrieus turbine* does not need much material relative to the power produced.

How can such a symmetric turbine start to revolve? Well, it can't, it needs a starter motor. But as soon as it starts to revolve, the wind will immediately take over, since the wind and the revolution speed together form a resulting wind that creates lift in the direction of revolution.

The swept area on a Darrieus turbine is $A = \frac{2}{3} D^2$. It has a narrow range of tip speed ratios around 6 and a power coefficient C_p just above 0.3. Several Darrieus turbines have been built, from a large MW prototype in Canada, through commercial turbines in the 150kW range, to small models of a few kW.

A *giromill* is a turbine with two or more straight vertical blades that form an H and are connected to a vertical axis.

A mechanical device that changes the blade angle during rotation can increase efficiency. This device makes it self-starting, but the blade angles have to be adjusted in relation to the wind direction to start. The swept area of a giromill is the height times the diameter: $H \times D$. The strong load on the blades' points of attachment and the bending moments are weak points of this design. Most prototypes have crashed and there are no large giromills manufactured commercially.

A giromill is more efficient than a Darrieus turbine and has a wider range of tip speed ratios, but it is not as efficient as a horizontal axis turbine with high tip speed ratio.

8
The Wind Turbine Rotor

A wind turbine rotor consists of rotor blades mounted on a hub. Most commercial wind turbines have a three-bladed rotor. There are, however, turbines with two blades, and in fact also turbines with only one single blade. The advantage with fewer blades is that the weight of the rotor and also of many other components of the turbine will decrease. The share of the power in the wind that can be converted will decrease with fewer blades, but from an efficiency point of view the differences are negligible, or at least easy to compensate by increasing the length of the rotor blade a bit.

On three-bladed turbines the connection between hub and blades is rigid. On a turbine with two blades or one single blade, they can be mounted so that they are flexible in the vertical plane. On a so-called teetering hub, the two blades can teeter a few degrees across the hub, which reduces loads on the turbine (see Figures 8.1 and 8.2).

The tip speed ratio of the rotor

To utilize the power in the wind in an efficient way the rotor has to have suitable rotational speed relative to its size (rotor diameter) and the wind speed. In other words it has to have an efficient tip speed ratio (see Chapter 6). The tip speed ratio of a turbine depends on the number of blades – fewer blades means the tip speed ratio should increase. This means that for turbines with the same rotor diameter, one-bladed turbines need a higher rotational speed than two-bladed turbines, which in turn need higher rpm than three-bladed turbines, and so on (see Figure 8.3).

There is also a relationship between tip speed ratio, rotational speed and the size of the rotor. The tip speed ratio is the ratio between the speed at the tip of the rotor blade and the undisturbed wind speed, given by $\lambda = w_{tip}/w_o$. At a given rotational speed the tip speed increases with the length of the blade.

The rotational speed is usually signified by n and with the unit rpm, revolutions per minute. The tip speed is given in the same unit as the wind speed, metres per second (m/s). Besides the rotational speed, the tip speed also depends on the radius of the rotor and is calculated using the following formula:

Figure 8.1 *Wind turbine with two rotor blades*

The wind turbine NWP 1000 from the Swedish company Nordic Wind Power, recently renamed Delta Wind, has a two-bladed rotor with a teetering hub. With such a 'soft' design concept the whole turbine can become slimmer and cheaper.

Source: Tore Wizelius

Figure 8.2 *Wind turbine with one rotor blade*

These turbines from the Italian manufacturer Riva Calzoni have a one-blade downwind rotor. The blade, which is balanced by a counterweight, is flexible in the hub and moves backwards when the wind speed increases, which reduces loads on the turbine.

Source: Riva Calzoni

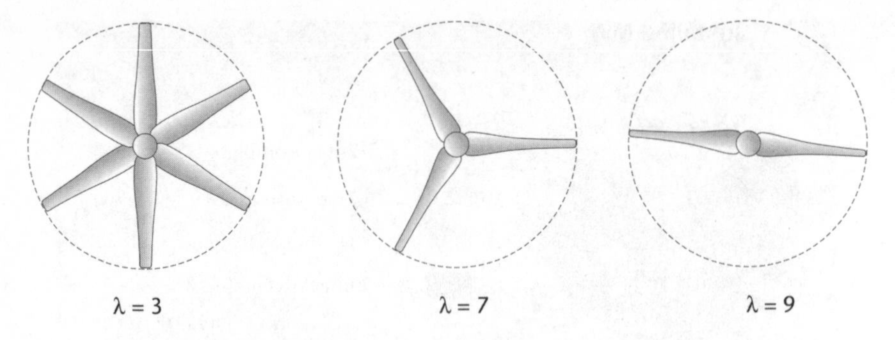

Figure 8.3 *Relation between number of blades and tip speed ratio*

Wind turbines with six, three and two blades respectively, and their tip speed ratios.

Source: Typoform/Södergård (1990)

$$w_{\text{tip}} = \frac{n 2 \pi R}{60} \text{m/s},$$

where R is the radius of the rotor.

The tip speed for a rotor with a 10-metre radius and a revolution speed of 30rpm is

$$w_{\text{tip}} = \frac{30 \times 2\pi \times 10\text{m}}{60\text{s}} \approx 30\,\text{m/s}.$$

If the radius is increased to 20m the tip speed will increase to ≈ 60m/s.

To keep the tip speed at 30m/s when the rotor radius is increased to 20m, the rotational speed can be reduced to 15rpm.

When the blades of a wind turbine rotate, the speed at the tip of the blade is higher than at the root or in the middle part of the blade. For a wind turbine with a 20-metre radius and a rotational speed of 30rpm the tip speed is 60m/s but the speed at the middle of the blade is only 30m/s.

Since the speed of the blade segments increases from the root to the tip, the apparent wind direction will change as well. When you move from root to tip, the apparent wind direction will move towards the vertical plane. Thus to get the same angle of attack along the whole blade, it has to be twisted (see Box 8.1).

Lift and circumferential force

The properties of the rotor also depend on the aerofoil used on the blades. In the early days of wind power (in the 1970s and 1980s), aerofoils developed for aeroplanes were used, mainly so-called NACA-aerofoils, but since the 1990s aero-

BOX 8.1 BLADE TWIST

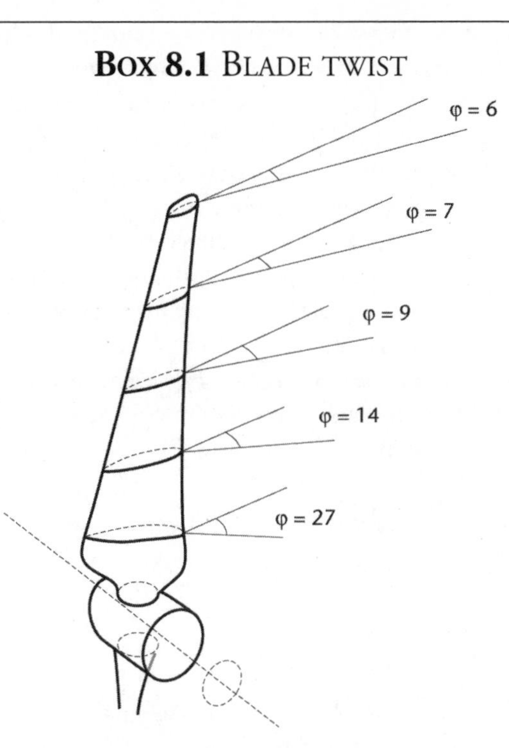

Figure 8.4 *Apparent wind direction along a rotor blade with a wind speed of 9m/s*

Source: Södergård (1990)

$$w_o = 9\text{m/s}$$

$\frac{2}{3} w_o = 6\text{m/s}$ (the undisturbed wind will decrease to $\frac{2}{3}$ just in front of the rotor disc)

$$w_{tip} = 60\text{m/s} \quad \varphi = 6°$$
$$w_{0.8R} = 48\text{m/s} \quad \varphi = 7°$$
$$w_{0.6R} = 36\text{m/s} \quad \varphi = 9°$$
$$w_{0.4R} = 26\text{m/s} \quad \varphi = 14°$$
$$w_{0.2R} = 12\text{m/s} \quad \varphi = 27°$$
$$\varphi = \alpha + \beta,$$

where φ = the angle of the apparent wind to the vertical plane,
α = angle of attack, and
β = blade angle (pitch)
(for definitions, see Chapter 6).

By twisting the blade (the blade chord's angle to the plane of rotation, β) so that the angle decreases towards the tip, the angle of attack α can be kept constant (for a given wind speed): $\beta = \varphi - \alpha$.

foils developed especially for wind turbines have been used. One profile can have several different thicknesses. The last two digits in the type number of an aerofoil give the relative thickness (thickness in relation to width) in per cent; thus the maximum thickness of NACA4412 is 12 per cent of the width.

On a wind turbine the lift from the rotor blades is utilized to make it revolve, but the circumferential force is not the same as the lift force. The lift is always applied perpendicular to the apparent wind direction; the blade has a certain (constant or variable) angle to the plane of rotation; and the aerofoil also offers some friction, drag (***D***), which is applied in the apparent wind direction. From these forces we get a circumferential force $\boldsymbol{F}_{circ}$ in the plane of rotation, and a force perpendicular to the plane of rotation, thrust $\boldsymbol{F}_{thrust}$. The circumferential force that propels the rotor looks quite small in relation to $\boldsymbol{F}_{thrust}$, but the power is large since the rotational speed is very high (see Figure 8.5).

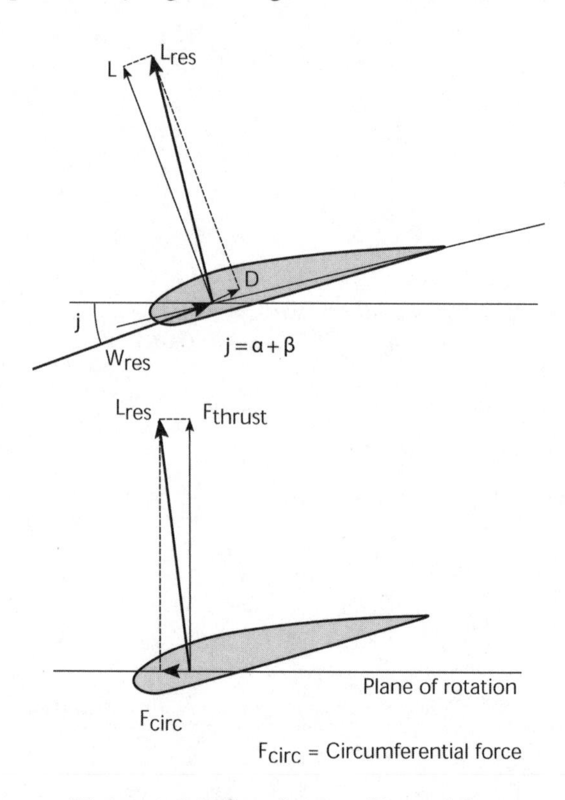

Figure 8.5 *Lift and circumferential force*

When a stream of air passes an aerofoil, drag ***D*** is created in the apparent wind direction ($\boldsymbol{W}_{res}$) and lift ***L*** perpendicular to this. These two forces have a resultant $\boldsymbol{L}_{res}$ (upper picture). $\boldsymbol{L}_{res}$ is divided into two forces: $\boldsymbol{F}_{circ}$ ('useful force', which is applied in the plane of rotation and makes the rotor revolve) and $\boldsymbol{F}_{thrust}$ ('useless force', which is applied perpendicular to the plane of rotation). This force is absorbed by the main shaft, main bearing and tower (lower picture).

Source: Typoform/Claesson (1989)

The aerofoils of a wind turbine should create a strong circumferential force and also have other properties suitable for wind turbines. The properties of an aerofoil can be read from the following diagrams, which show the test results of profiles from a wind tunnel. The first shows the relationship between the angle of attack and lift (C_L), the second, a gliding ratio diagram, shows the relationship between the lift coefficient C_L and the drag coefficient C_D (see Figure 8.6).

The direction of the wind that passes the blades will vary along the blade; the angle in relation to the plane of rotation will decrease towards the blade tip. The apparent wind direction will also change every time the undisturbed wind speed changes. If the blade angle and rotational speed are constant, the angle of attack and therefore also lift, drag and gliding ratio will change continuously between different parts of the blade (see Figure 8.7).

Up to the end of the 1990s most wind turbines had a constant rotational speed, which for turbines with 1MW nominal power and about 50m rotor diameter would be around 25rpm. Some turbines used two different fixed rotational speeds: a lower one for low wind speeds, when a small generator was used, and a higher one for stronger winds, when the large generator cut in. In this way

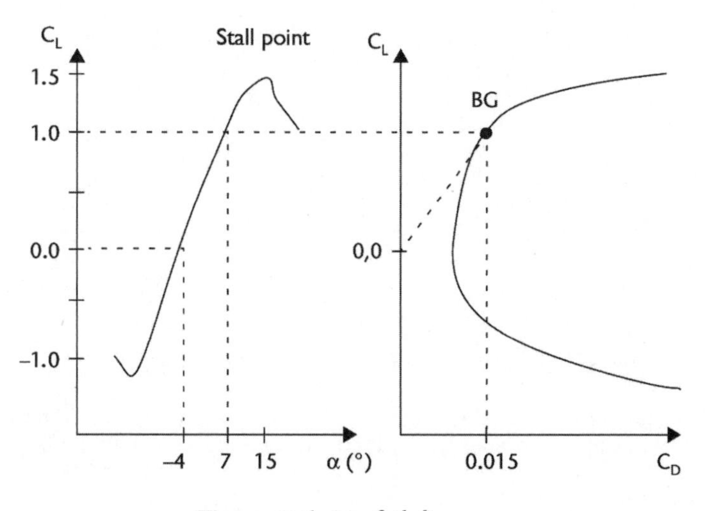

Figure 8.6 *Aerofoil diagrams*

The left diagram shows the relationship between lift and angle of attack. The aerofoil that the diagram describes starts to have lift with an angle of attack of −4° and reaches a maximum at 15°, after which it decreases. With such a large angle of attack, the airstream cannot stick to the surface of the aerofoil, so turbulence is created. The airflow separates and the aerofoil begins to stall.

The right diagram shows the relation between C_L (the lift coefficient) and C_D (the drag coefficient). These are standardized aerodynamic coefficients that make it possible to calculate lift and drag for aerofoils of different sizes. The best gliding ratio is found at the tangential point of a line from 0 on the C_L axis and the graph. The aerofoil is most efficient with an angle of attack of 7° and has a safe distance to the point where stall occurs at $\alpha > 15°$.

Source: Claesson (1989)

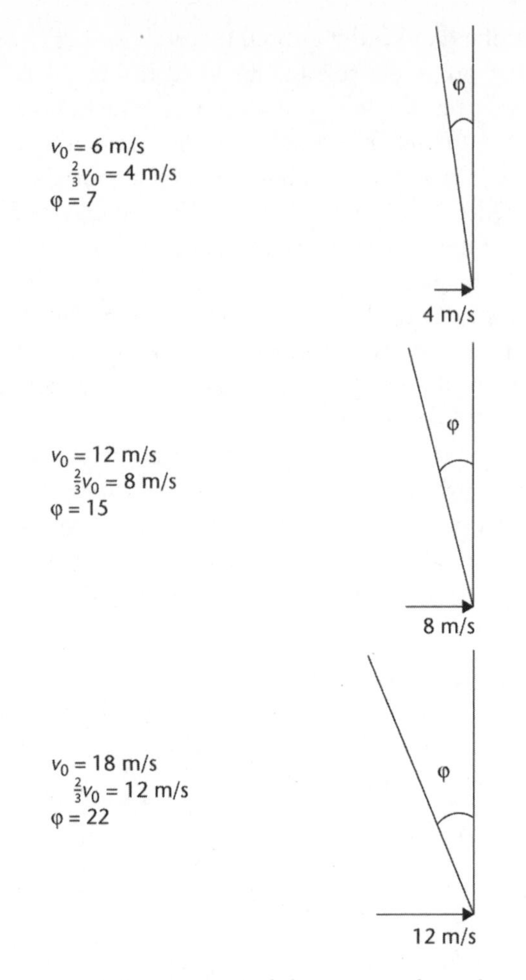

Figure 8.7 *Apparent wind direction and wind speed*

Apparent wind direction over a segment of rotor blade with a rotational speed of 30m/s. On wind turbines with constant rotational speed and blade angle, the angle between the plane of rotation and the apparent wind speed will increase when the wind speed increases. The angle of attack will increase simultaneously.

Source: Tore Wizelius

the rotor can operate close to the optimum tip speed ratio at all wind speeds. In the ideal case, however, it is the tip speed ratio that should be fixed, and not the rotational speed, in other words the rotational speed should increase with the wind speed. However, if the rotational speed of the rotor varies, the generator will also get a variable speed and the voltage and frequency will vary as well.

Some manufacturers used variable rotational speed combined with power electronics (rectifier and inverter) back in the 1980s, but in recent years more and more manufacturers have changed the design concept from fixed to variable rotational speed. One driving factor behind this has been that the prices for power electronics have decreased quite substantially. Another is the demand from grid operators on power quality from wind turbines.

Power control

The wind that makes the rotor revolve also pushes the whole turbine backwards. If this thrust force gets too strong, the components in the turbine will be overloaded and can break or the whole turbine can even be pushed over. Strong winds can break the trunks of trees and lift a roof off a house. Wind turbines stand where they have been installed and have to endure the worst weather conditions that can appear at the site.

Power control means that the turbine limits the amount of (or share of) the available power in the wind when the wind speed reaches a preset value, the rated wind speed. This is usually set to a value between 12 and 16m/s. The rated wind speed differs between different models and manufacturers. It can also be adapted to the wind conditions at a specific site by adapting the rotor diameter to the generator's rated power. When the wind speed increases to storm force, the rotor will be stopped and parked to protect the turbine from damage. This so-called *cut-out wind speed* is set to about 25m/s for most turbines.

There are basically two different methods for power control on large wind turbines: *pitch* and *stall* control. Turbines with *pitch control* have rotor blades that can be turned around their longitude axis from the hub. By turning the blades the angle of attack can be adapted to the wind speed. When the wind speed increases above the rated wind speed, the blades are turned so that the power that is extracted from the wind is reduced. In this way the power extracted from the wind can be kept constant, at the rated power, when the wind speed rises above the rated wind speed; the power curve will thus level out. When the cut-out wind speed is reached, the blades are turned out of the wind so that the wind can blow through the rotor without creating any lift and the rotor stops rotating. The blades are *feathered.*

Turbines with *stall control* have a blade profile that creates eddies (turbulence) on the upper side of the blade when the wind speed increases beyond the nominal wind speed. The lift decreases and the drag increases. In this way the power that is extracted from the wind can be kept close to the nominal power of the turbine even when the wind speed is higher than the rated wind speed. However, it is difficult to manufacture a rotor blade that will stall exactly as much as desired at a specific wind speed to keep the power constant at the rated power in strong winds (in fact this is impossible taking into account variations in air density with variations in temperature and air pressure). The stall usually increases gradually: it starts at 8–9m/s and thereafter increases so that the power is kept as close as possible to the rated power of the turbine when the wind speed increases above the rated wind speed. When the cut-out wind speed is reached, aerodynamic brakes are activated to stop the rotor. This brake in most cases is set at the blade tips: the tip section of the blades is twisted to stop the rotation.

Stall control has been developed into so-called *active stall*. In active stall systems the rotor blades can be turned along the longitude axis, just like pitch controlled turbines. To reduce the extraction of power when the wind speed increases, pitch controlled turbines turn the blades so that the wind will pass the turbines more easily by decreasing the angle of attack. Turbines with active stall control turn the blades the other way, to increase the angle of attack and consequently the stall.

Wind turbines with variable rotational speed don't need to regulate the blade angle, since rotational speed increases with wind speed to keep the angle of attack constant (and optimized). When a certain rotational speed and power output has been reached, this kind of turbine also has to control the power by pitching the rotor blades.

Box 8.2 Power control

The windmills of the past were run manually by the miller. When winds were too strong, the windmill was not used. When they were out of use, the rotor blades did not have their sails on (they were made of sailcloth or wood boards), and the rotor was parked with a chain. If unexpected strong winds appeared when the windmills were running, some windmills could reduce the coverage of the rotor while running, by a safety measure that stripped the sails off the rotor. Others had self-regulating sails with shutters, so-called *patent sails*. Most windmills, however, had to be stopped with mechanical brakes, or turned out of the wind manually. This was not easy, and many windmills caught fire from sparks from the brakes when the rotor got out of control. This is why there are so few windmills left today: most of them have vanished into smoke.

Wind wheels had other methods for power control. A large oblique vane was a simple method. When wind speed increased, the pressure from the wind on the vane increased so much that the rotor was turned out of the wind. On the Excenter turbine the nacelle (generator, etc.) was mounted eccentrically on the tower, so it started to turn out of the wind when the winds pressure on the rotor exceeded the rated wind speed. On other small turbines, the rotor and nacelle are turned over backwards, over a hinge in the back of the tower top, so the rotor goes into a helicopter position. The power decreases when the rotor swept area is reduced.

The rotor blades can also be used for power control, either by using an aerofoil that stalls when the wind speed gets high enough or by making the blades turn along the longitude axis. Many more or less complicated methods for this have been developed. A centrifugal regulator consists of a metal bar with a weight, attached to the blade close to the hub. The blades are turnable, but are held in position by a spring. When the rotor revolves, these weights apply a force to the blades, created centrifugally. This force increases with the rotational speed, and when the force gets stronger than the force of the springs, the blades start to turn and reduce the power extracted from the wind.

Rotor blade design

In practice rotor blades must have many different properties and have to be designed to stand the strong loads that they will face. Therefore it is not suitable or possible to use ideal blades that give maximum efficiency. Most blades are designed with three different aerofoils: a very thick one close to the hub, an aerofoil with average thickness in the middle part of the blade, and a very thin one close to the tip. The part closest to the hub has to be thick and strong to make a solid connection to the hub. This part of the rotor blade is almost round, like a pipe. This innermost part of the rotor has a small area and does not contribute much to power production.

The outer part of the blade often also has a less than optimum blade width, since you gain more by increasing the swept area than you lose by using a less efficient blade width, which needs more material. It is the weight that is the decisive factor in how long the blades can become.

Stall-controlled turbines with a fixed blade angle would be very hard to get started in slow winds if they were twisted to attain optimum efficiency at their constant rotational speed. A part of the blade is therefore designed with a blade angle that will give enough lift to get the rotor to start revolving at a wind speed of 3–5m/s.

When the wind speed approaches the rated wind speed on a stall-controlled turbine, and the power has to be controlled by stall, the revolutions will become jerky if the wind stalls along the whole blade at the same time. By choosing a twist and a combination of aerofoils, so the stall will vary on different segments of the blade, this can be avoided. The stall starts close to the root of the blade and spreads over an ever-larger part of the blade as the wind speed increases, so the power levels out when the rated power/wind speed is reached.

Pitch controlled turbines can turn their blades to get the rotor to start to revolve after a calm. They could in theory also vary the blade angle continuously to adapt it to the prevailing wind speed. This has, however, not proved a reasonable strategy, since the wind speed changes so frequently and rapidly. It would be difficult to regulate fast enough, and components would become worn out very quickly. (On new and very large turbines this technology might prove practicable.) The pitch control is mainly used to control the power when the wind speed is above the rated wind speed. Pitch controlled turbines can use other aerofoils, since they don't depend on stall. Turbines with variable speed can keep the blade angle constant since the angle of attack will be constant as well.

Several manufacturers of stall-controlled turbines have developed active stall. The advantage of this is that the power can be controlled better. The power in the wind depends not only on the wind speed, but also on the air density, which varies with air pressure and temperature. In conditions with the same wind speed, the wind will contain more power in the winter when the temperature is –10°C than in the summer when it is 20°C. In the winter a stall-controlled turbine with nominal power of 1MW can thus produce 1.1MW, which is not good for the generator. In the summer the opposite applies, and the turbine can't achieve its full capacity. By regulating the blade angle the turbine can be adjusted to optimize the turbine in all weather conditions. The power can also be controlled when the wind speed is stronger than the rated wind speed, so the power output will stay at the rated power level.

Wind turbines with pitch control have a pitch mechanism installed in the hub. Some manufacturers use a hydraulic system for this, with a pump in the nacelle and a piston that passes through the main shaft to the hub, where the movement is transferred mechanically to the blades. Other manufacturers use an electric system, with an electric motor connected to each of the blades. The latest development is to have individual pitch on the blades, so they can be turned independently of each other. On very large turbines the hydraulic system can be installed in the hub instead of in the nacelle.

If a rotor runs without a load, it will accelerate very quickly to its maximum tip speed ratio. A three-bladed turbine is most efficient at a tip speed ratio of 6–7, but without a load the tip speed ratio rapidly increases to 18 before the turbine loses power. A rotor blade on a turbine with 50m rotor diameter will then have a tip speed of 180m/s (650km/h) at a wind speed of 10m/s. This is usually more than any blade will bear, so it is liable to break.

The load on a turbine that is generating electric power is in the generator. If the turbine is disconnected from the grid, during a power cut for example, the load disappears and the rotor runs free. Therefore all turbines must have an aerodynamic brake. If the blades can be turned, they can simply be turned so that the rotor stops. When the blades are turned to a horizontal position, called 'feathering the blades', they lose their lift and are stopped by the friction from the air. Stall-controlled turbines have blade tips that can be twisted. The outer parts of the blades are turned into a perpendicular position in relation the rest of the blade. Inside the nacelle there is also a mechanical brake which can be applied when the rotor speed has decreased and which can be used as an emergency brake if the aerodynamic brakes fail.

Rotor blades are exposed to great strain and stress. Since the wind is constantly changing, the blades are exposed to millions of load changes and have to be manufactured with material that can stand such stress and avoid fatigue. Steel and aluminium can't do this cost-effectively, so most rotor blades are manufactured using glass fibre or epoxy. Wood is also a material that has good fatigue properties and there are blades made of laminated wood with a plastic coating. The blades are built around a load bearing axis. The surface, made of sheets of glass fibre or epoxy, is mounted around the axis to create an aerofoil. The relative weight of the blades (kg/m^2 swept area) has been halved since the early 1980s, from 3 to 1.5kg/m^2 swept area. And with carbon fibre and glass fibre reinforced epoxy the weight could be reduced even further, down to 0.5–0.7kg/m^2.

As the turbines have grown in size, rotor blades have been made more elastic. Some of the loads can thus be absorbed directly by the blades instead of by the nacelle and tower. With an upwind rotor, however, there is a limit to the elasticity that can be built into the blades, as obviously they can never be allowed to hit the tower.

Wind turbines that are built for an arctic climate may also need a de-icing system for their blades. With cold rain and fog, when the rotor is idle, a crust of ice will quickly form; the shape of the aerofoil will thus change and so will the aerodynamic properties of the rotor. Ice can also create imbalance on the rotor, in which case the turbine will not be able to start, as the control system will prevent unbalanced running. Thus a turbine can lose a lot of valuable production during the winter. To avoid this, de-icing systems that melt the ice can be used. One system uses electric heating of the blades and is governed by an ice detector on the nacelle. The anemometer and wind vane should also be heated to avoid icing in arctic climates, since they give important input to the wind turbine control system.

9
Nacelle, Tower and Foundation

The unit mounted on top of a wind turbine tower is called the nacelle, gondola or machine cabin. Inside the nacelle there is the gearbox, generator and other mechanical and electrical components. Most large grid-connected wind turbines have conical steel towers. Smaller turbines can have a lattice tower or guyed mast. To make turbines firmly rooted in the ground, so that they will not be turned over by strong winds, turbines are mounted on foundations of reinforced concrete. If the bedrock is solid and stable they can be bolted to the rock.

Nacelle

The nacelle of horizontal axis turbines contains a bedplate on which the components are mounted. There is a main shaft with main bearings, a generator, and a yaw motor that turns the nacelle and rotor into the wind. There can be several other components as well; these vary depending on the model and design concept used by the manufacturing company.

A wind turbine of the so-called Danish standard concept, which has been used since the early 1980s (three-bladed upwind rotor with fixed rotational speed and an asynchronous generator) contains the main shaft (the shaft that is turned by the rotor) and main bearings, an asynchronous generator and a gearbox that will increase the rotational speed of the rotor to the 1010 or 1515rpm that the generator demands to produce electric power. There is also a yaw motor and a disc brake used for emergency stops and parking. On top of the nacelle an anemometer and a wind vane are mounted; these are connected to the control system of the turbine (see Figure 9.1).

The main shaft protrudes through the front of the nacelle. A rotor hub made of a steel casting is mounted at the end of the shaft. The hub is covered by a nose cone that protects the hub and reduces the turbulence in front of the rotor. Wind turbines with pitch or active stall control also have bearings for the blades and mechanical or electrical equipment to adjust the blade angles. Turbines that use a hydraulic system to rotate the blades have a hydraulic pump in the nacelle connected to a piston that passes through (inside) the main shaft out to the hub.

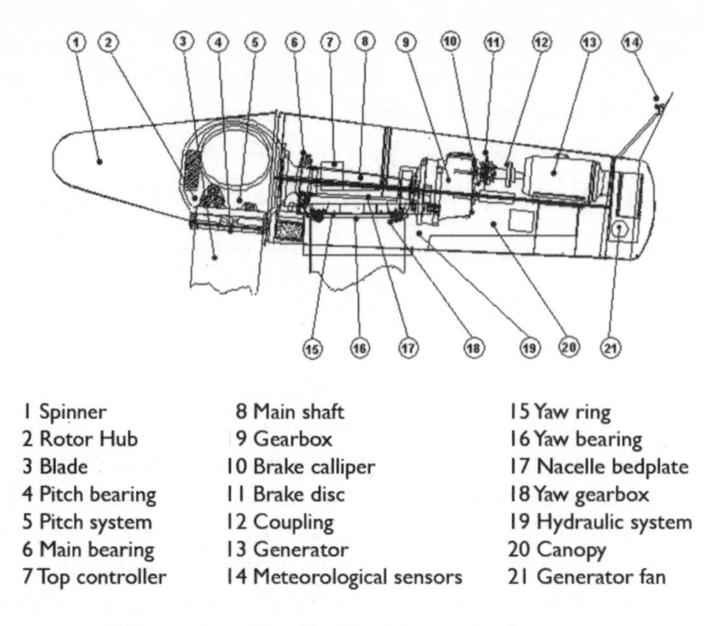

Figure 9.1 *Nacelle: Danish standard concept*

The nacelle of a Siemens 1.3MW turbine.

Source: www.powergeneration.siemens.com

The purpose of the gearbox is to increase the low speed of the main shaft to the speed that the generator demands – 1010rpm for a six-pole and 1515rpm for a four-pole generator. Since a large wind turbine has a rotational speed of 15–30rpm, a significant step-up is necessary; this has to be done in several steps. Most wind turbines therefore use a three-step gearbox. Gearboxes on large turbines also need efficient lubrication and cooling and thus need an oil pump and oil cooling system as well.

Most wind turbines use so-called *asynchronous generators*. The generator size is specified by its nominal power, which will be attained at the nominal wind speed. When the wind speed is lower, the generator will produce less power. Many wind turbine models have two different generators, or a so-called *double wound generator* (which can alternate between four and six poles, in other words like two generators in one). The smaller generator is utilized for low wind speeds and the larger for high wind speeds. There are also wind turbines that have a *multi-pole synchronous generator*, which can produce electric power at low revolution speed and does not need a gearbox. The nacelle of such a turbine contains very few components (see Figure 9.2 and Figure 10.7 in Chapter 10).

A new and very interesting concept is the so-called hybrid, which combines a multi-pole synchronous generator with low revolution speed with a robust one-step planet gearbox.

Wind turbines should have two independent brake systems, an aerodynamic brake (described in Chapter 8) and a mechanical brake in the nacelle. The me-

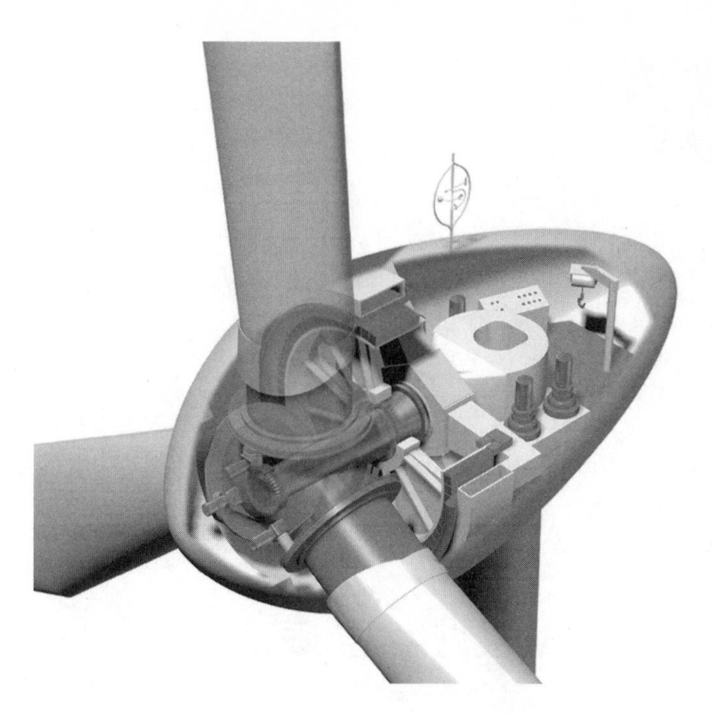

Figure 9.2 *Direct drive wind turbine*

Nacelle of an Enercon E-48 turbine. Wind turbines manufactured by Enercon have a very large multi-pole ring generator that is connected directly to the rotor. Inside the hub there are three electric motors for pitch control of the rotor blades, and in the nacelle there are yaw motors to align the turbine to the wind direction.

Source: www.enercon.de

chanical brake is mainly used as a parking brake when service and maintenance is performed in the nacelle, but it should have enough strength to stop the rotor if the aerodynamic brakes fail. A disc brake, mounted on the fast running shaft that connects the gearbox with the generator, is used for this purpose.

Yaw control

To utilize the wind efficiently, the rotor should be perpendicular to the wind direction. When post mills where in use, the miller simply checked the wind direction, and turned the windmill towards the wind by hand or with the help of oxen or a winch. But even in the early days of wind power the windmill engineers were ingenious. They developed the so-called Dutch windmill. The top of the mill, the cap, was disconnected from the tower so it could be turned on a slide bearing to bring the sails into the wind. A wind wheel was mounted perpendicular to the rotor. When the wind direction changed so it hit the windmill rotor sideways, this

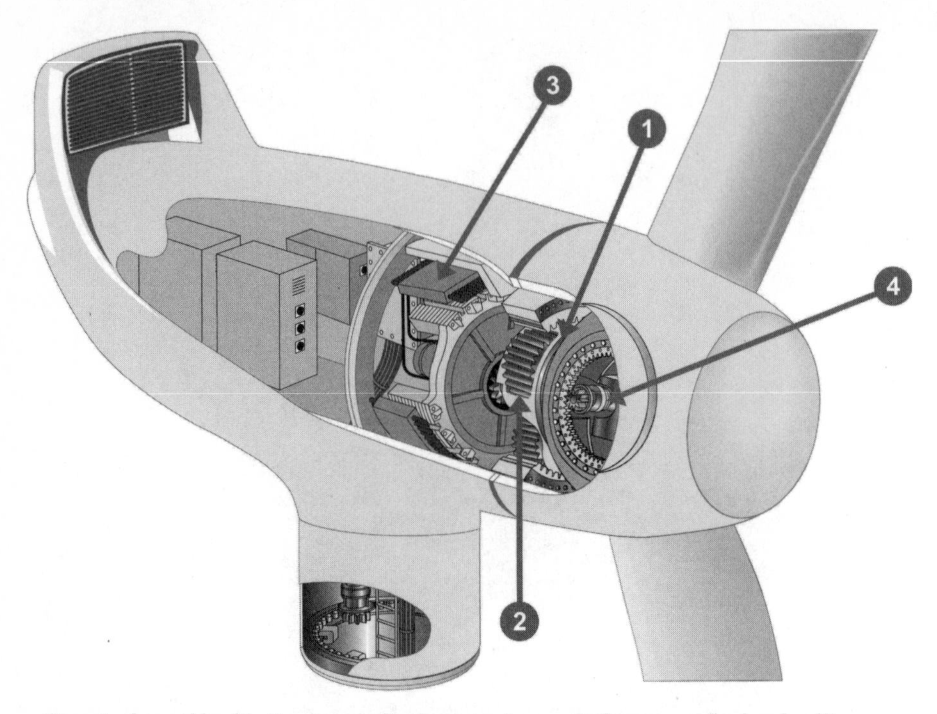

The rotor is combined to the power unit using a custom-made three-row roller bearing (1).
The roller bearing transfers the rotor loads directly to the main casing past the planetary gear and generator.
The single-stage planetary gear (2) increases the rotating speed from 8-25 rpm to 44-146 rpm.
The low speed permanent magnet generator (3) produces the electricity.
The rotational speed of rotor is controlled by three independent electric pitches (4).

Figure 9.3 *Hybrid concept*

Nacelle of the WinWind turbine, manufactured in Finland. It has a robust one-step planet gearbox connected to a multi-pole synchronous generator.

Source: WinWind

wind wheel started to revolve. And the wind wheel was connected to a cogwheel that turned the cap towards the wind. When it was back in its perpendicular position towards the wind, the wind wheel stopped. This robust mechanical yaw system is still used on some smaller turbine models.

Larger turbines use yaw motors that are controlled by a wind vane. When the wind changes a specified number of degrees, and this new wind direction lasts a specified time, the control system sends an order to the yaw motor, which begins to turn the nacelle back into alignment with the wind. If the nacelle has made several revolutions in the same direction, the cable from the nacelle down to the ground has to be rewound; otherwise it will be twisted off. After three revolutions (on most models) the turbine will be stopped by the control system and the yaw motors turn the nacelle in the opposite direction to rewind the cables.

Towers

Most manufacturers of large turbines use conical tube towers made of steel, painted in white or grey, which are wider at the base than the top. In the 1980s, when the turbines were just 30m high, these could be welded in one piece. On large turbines with hub heights of 40–120m, the towers are manufactured in sections, which are bolted or welded together when they are mounted. The towers have a door at ground level, and the control system, displays and some electric equipment are installed inside the tower. There is also a ladder up to the nacelle on the inside of the tower.

Some manufacturers offer concrete towers as an option. These have a smaller diameter and the ladder is mounted on the outside (on small and medium-sized turbines). The control system and electric equipment are installed in a separate control room in a container next to the tower.

Lattice towers were also often used on small wind turbine models in the 1980s. Nowadays only a few manufacturers use them, but they are still common for smaller turbines. Lattice towers have many advantages: they use less material and have lower weight and price. Another advantage is that the winds can pass through the tower, which reduces loads on the turbine. Most manufacturers, however, prefer to use steel tube towers (although there are some MW-turbines mounted on lattice towers) for practical and also aesthetical reasons. In Denmark, for example, it is not permitted to use lattice towers for large turbines. Small turbines also often use guyed masts as towers.

Foundations

The foundations that a wind turbine is mounted on have two functions – to carry the weight of the turbine (to prevent it from sinking into the ground) and to act as a counterweight to prevent the turbine from tipping over. The design and weight of the foundations has to be adapted not only to the size of the turbine but also to the soil properties at the specific site where it will be installed.

On ordinary soil a 2–3m deep cavity is dug in the ground, forming a square with a 7–12m side. The dimensions depend on the size of the turbine, its weight and hub height, and the ground conditions. On waterlogged ground the foundation has to be bigger to compensate for the lifting force from the groundwater.

When the bottom of the cavity has been levelled, reinforcement bars are mounted in layers separated by distance pieces. In the centre a pillar is formed up to ground level that will be used as the mounting base for the tower. After that the concrete foundations are completed. The concrete then has to harden for a month before it is covered by filling material and the tower can be mounted (see Figure 9.4).

Figure 9.4 *Gravity foundation*

Foundation for a 1MW turbine. When the concrete has hardened for a month, the foundation will be covered by earth to restore the ground.

Source: Nordic Wind Power

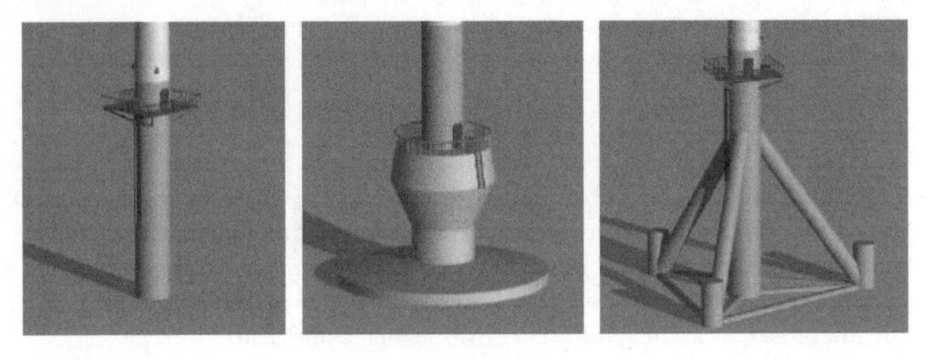

Figure 9.5 *Offshore foundations*

There are three different kinds of foundations for offshore wind turbines: monopiles (left), gravity foundations (middle) and tripods (right).

Source: Henderson (2002)

If the wind turbine is to be installed on rock, bolts in the rock can anchor the foundations. A number of deep holes are drilled in the rock, long bolts are inserted into the holes and expanding concrete is injected that fixes the bolts in the rock. Each of the bolts has to stand a tractive force of 30 tonnes (for a 600kW turbine). A mounting base for the tower is then founded and anchored to the bolts.

For offshore wind turbines many different types of foundations are under development for use in deeper waters, mainly different types of tripods (three-legged foundations). So far two types have been used: *gravity foundations* and *monopiles*. The type of foundation that is most suitable depends on the character of the seabed. Gravity foundations are manufactured at a shipyard and are made of reinforced concrete. The seabed is levelled out and the foundations are towed to the site, filled with some heavy material and submerged.

A monopile is simply an elongation of the turbine tower. The monopile can be inserted in a hole that has been drilled in the bed or driven down by a pile driver. There are other kinds of offshore foundations as well, but so far only these two options have been used for large offshore wind farms (see Figure 9.5).

10
Electrical and Control Systems

Most modern wind turbines are used to convert the kinetic power in the wind to electrical power. The rotor transfers the kinetic power in the wind to a revolving shaft that drives a generator that generates electric power. A generator is made of a revolving part, the rotor, and a stationary part, the stator. The rotor in the generator has a magnetic field, which is created either by permanent magnets or electromagnets. When the wind turbine starts to revolve, it thus creates a rotating magnetic field. When this magnetic field passes the stationary coils, an electric current is induced in them and this current can be fed into the power grid.

Most generators generate alternating current, AC. This means that current and voltage will change directions several times during each revolution of the rotor. The frequency of the AC current, the number of periods (from 0 to + to 0 to – to 0, in a sine curve), depends on the rotational speed of the generator. To get a constant frequency from the generator, the rotational speed of the wind turbine rotor has to be fixed. If the rotational speed varies, the frequency and voltage will vary as well.

Electric systems in wind turbines

Wind turbines with generators that are directly connected to the power grid have a rotational speed that corresponds to the grid frequency. In Europe the frequency is 50Hz, 50 cycles per second (in the US the frequency is 60Hz). A simple generator with only two poles (N and S) would then need a rotational speed of 50 × 60 seconds = 3000rpm to give 50Hz. The number of poles in a generator can, however, be increased to give more than one cycle during one revolution – a generator with four poles gives 50Hz at 1500rpm. The relationship between rotational speed and number of poles is calculated as $n = 6000/p$, where p is the number of poles. Most mass produced generators have four or six poles.

In the power grid a three-phase AC is used, which means that three alternating currents run parallel with each other. These three currents are displaced a third of a period to each other. Thus the generator has to generate three separate AC currents and needs a set of poles for each of them.

Some manufacturers use so-called ring generators. These have a large number of poles, not 4 or 6 but 64 or 96 or some other number of poles, depending on the size and usage of the generator. They also have a large diameter and can therefore be run with a low rotational speed (the peripheral speed increases with the diameter, just like the tip speed of the wind turbine rotor). By increasing the number of poles and the diameter, the rotational speed necessary to generate electric power in a reasonably efficient way can be reduced to the speed of the turbine rotor. With this design concept a gearbox is not needed.

Two basically different kinds of generators are used in wind turbines: synchronous and asynchronous. A synchronous generator can be connected to the grid, or work without grid connection (connected to a local grid, battery storage or local loads such as an electric water pump). An asynchronous generator has to be connected to a grid to function, since it is dependent on it to magnetize the rotor and it is governed by the grid frequency. When there is a perfect fit between the rotational speed of the generator and the grid frequency, the generator runs idle. To generate power, its rotational speed has to be asynchronous, in other words a little higher than the frequency. An asynchronous generator with a nominal rpm of 1000 has to be run at 1010rpm to produce full power (see Figure 10.1).

Asynchronous generators are installed in turbines with a fixed rotor speed. Actually, the rotational speed is not completely fixed – if the rotational speed is lower than the synchronous speed, the machine will give a positive torque; in other words it works like an electric motor. If the shaft is revolved (by the wind turbine rotor) a little faster than the synchronous rpm, it will start to function as a generator and feed power to grid. The normal operating range is from the synchronous rpm up to the nominal rpm, which is about one per cent higher (at

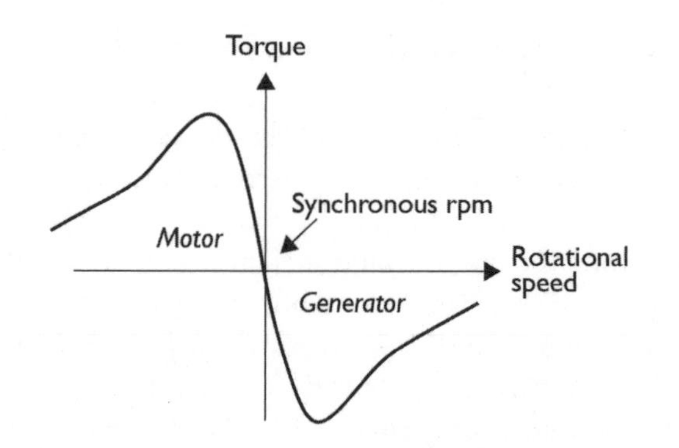

Figure 10.1 *The moment curve of an asynchronous machine*

The relationship between the torque of the shaft and the rotational speed for an asynchronous machine.

Source: Claesson (1989)

a torque of 100 per cent in the diagram). When the wind speed increases and the torque increases, the rotor tries to accelerate, but instead the force of the magnetic field in the rotor is increased, which results in increased power in the generator and more power fed to the grid.

The operating rpm deviation from the synchronous rpm is called *slip* and is expressed as percentage deviation from nominal rpm. This slip will be highest at the nominal power of the generator, when it is around 1 per cent. When the generator is working at full capacity, the torque on the rotor shaft cannot be allowed to increase. If the wind speed gets stronger some of the power that turns the rotor around must be diverted (or spilled) by some means of power control, pitch or stall. Otherwise the generator will be overloaded. If a situation with too much power generation lasts too long, the generator will crash (the windings will melt from the heat, for example).

During a shorter period the nominal power can be exceeded. This happens, for example, before the pitch control system has adapted the blade angle to a new higher wind speed, caused by gusts. During short intervals the generator can be run as motor as well, to avoid the turbine cutting in and out too often when the winds are close to the cut-in wind speed.

A synchronous generator does not need a grid connection to be able to produce power. It can be used for a local grid, or be directly connected to an electric water heater, where the electric resistors that create the heat can be connected stepwise and adapt the rotor rpm to the wind speed by changing the load. By doing this an efficient tip speed ratio can be maintained. The frequency will vary, but for a heating load this does not matter.

Relationship between wind speed and power

The size of generator is defined by its nominal power. On a wind turbine this power will be reached at the nominal wind speed (12–16m/s, depending on manufacturer, model, site, etc.) and higher wind speeds. At lower wind speeds the power is significantly lower. The relationship between wind speed and power is shown in *power curves* (see Figure 10.2).

At a certain maximum wind speed (in most cases 25m/s) the wind turbine will be stopped and disconnected from the grid. The loads on the turbines will be reduced when the turbine is stopped and parked and the rotor is not revolving. In this parked state, wind turbines will survive winds up to hurricane force (60m/s). The maximum wind speed that turbines are designed to withstand is called the *survival wind speed.* The lost production from periods with wind speeds above the cut-out wind speed, > 25m/s, are negligible, however, since these wind speeds are very rare and occur only for short periods.

The American Charles F. Brush built the first wind turbine that produced electric power in 1887–1888. It was operational for 20 years and was used for bat-

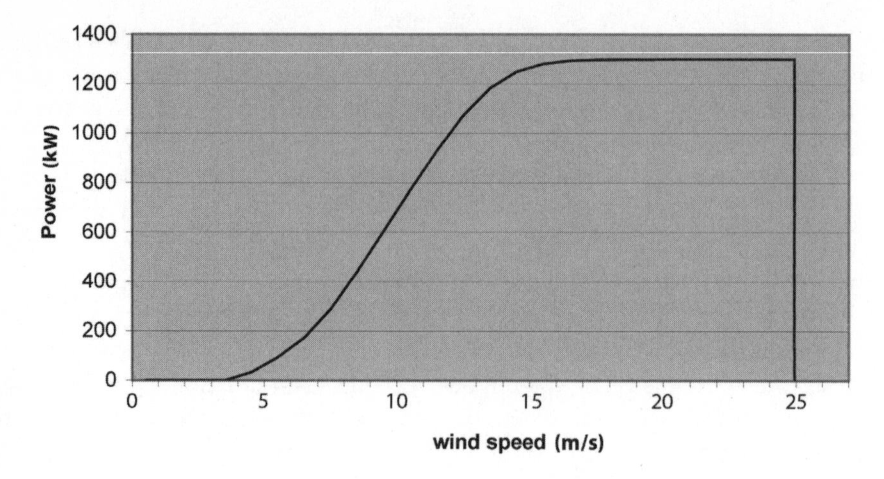

Figure 10.2 *Power curve for a wind turbine*

The power curve shows how much electrical power a wind turbine will produce at different wind speeds. This turbine starts to produce power at 4–5m/s and reaches the nominal power, 1300kW, at 16m/s. With higher wind speeds the power levels out, and at 25m/s, the cut-out wind speed, the turbine is stopped by the control system.

Source: Tore Wizelius

tery charging. In Denmark too wind turbines for electric power production were developed around the turn of the 20th century. These turbines were made for stand-alone systems and were used to charge batteries, heat water and run electric water pumps. The early turbines had DC-generators. In these kinds of generators the power is produced in the rotor and is transmitted from the generator by a revolving contact; a *commutator*. A serious drawback with DC generators is that the commutator wears out, which increases the need for servicing. These kinds of generators are therefore not used very much nowadays.

If an AC generator is used to charge batteries, the current first has to be converted to DC by a rectifier. For this kind of wind turbine, with a power rating from a few hundred watts to a few kW, there is a considerable niche market today. They are used as battery chargers on caravans and sailing yachts and in holiday cottages, and for lighthouses, telecommunications masts and offshore oil platforms in areas with no grid connection. The rotational speed on small turbines is so high that no gearbox is needed. Most of them have direct drive synchronous generators with permanent magnets.

Asynchronous generator with fixed speed

The first commercial wind turbines for grid connection that were installed during the late 1970s and the early 1980s used standard asynchronous generators that

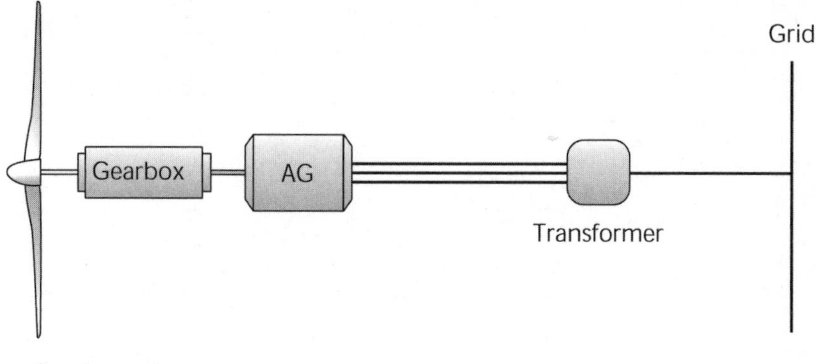

Figure 10.3 *Asynchronous generator with fixed speed*

The load in an asynchronous generator will increase in proportion to the power transmitted from the rotor, so the speed will remain fixed (with very minor variations). To increase the rotational speed to 1500rpm a gearbox is necessary.

Source: Typoform (after Tore Wizelius)

were available on the market and rotors with stall control. They had a fixed speed and used a very simple electrical system (see Figure 10.3).

This simple design caused some problems for grid operators, who claimed that the wind turbines had a negative impact on power quality, especially when the number of turbines increased. The impact when a wind turbine generator cut in to the grid was actually the same as an electric motor of the same size being started. When the turbines start and are connected to the grid, the generator needs reactive power to magnetize the rotor, and when the generator is connected to the grid a short but powerful current, a *spike*, sets in and the voltage drops for a split second. During operation the turbines produce active power, but they also consume some reactive power from the grid. This put unnecessary demand on the grid capacity.

Soft start generator

In those days the turbines were quite small, and the disturbances on the grid were within tolerable limits. However, additional equipment to reduce these problems was soon installed on the turbines: *capacitors* to reduce the reactive power consumption and so-called *soft start equipment*, a couple of thyristors to reduce the cut-in currents (see Figure 10.4).

When a generator runs on a partial load (for example a 500kW generator producing just 100kW) the efficiency of the generator will be considerably lower than when it is running on full power. With a fixed rotational speed that has been set to get a good tip speed ratio in relatively strong winds, the efficiency of the

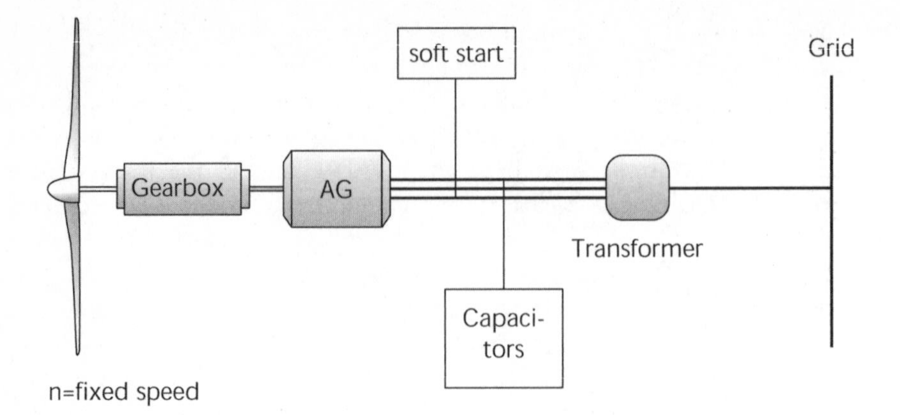

Figure 10.4 *Grid adapted asynchronous generator*

To eliminate spikes when the turbine is cut in to the grid, soft start equipment is used; to reduce the demand for reactive power, capacitors are installed.

Source: Typoform (after Tore Wizelius)

rotor will be comparatively low at low wind speeds, since the tip speed ratio will be too high.

Double generators

To utilize the power in the wind in a more efficient way, especially in so-called low wind areas (with low average wind speed), some manufacturers began to install two separate generators, or a so-called double wound generator (which works as two generators in one). A small generator with six poles and 1000rpm runs on full load in low wind speeds with a lower rotational speed (better tip speed ratio) on the rotor. When the wind speed increases over a certain limit (~ 7m/s), the large generator, with four poles and 1500rpm, takes over and the rotational speed increases to a higher fixed level (see Figure 10.5). By reducing the rotational speed at low wind speeds, the aerodynamic swish-noise from the rotor will also be reduced, which is an important advantage since the noise generated is the most important limiting factor for the siting of turbines in inhabited areas.

Variable speed

To get maximum efficiency in utilizing the power in the wind, the rotational speed of the wind turbine rotor has to be proportional to the wind speed. With a variable speed the tip speed ratio can be kept at the optimal level for all wind speeds. Turbines with a high tip speed ratio operate efficiently over a wide range of tip speed values; it can vary quite a lot around the optimal value and still main-

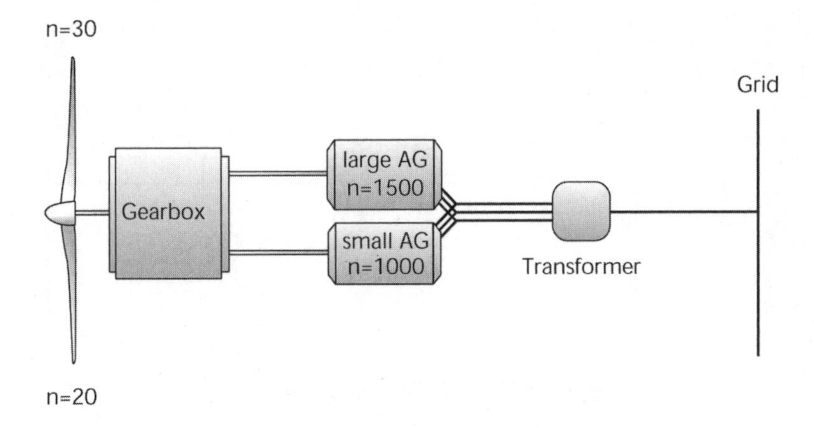

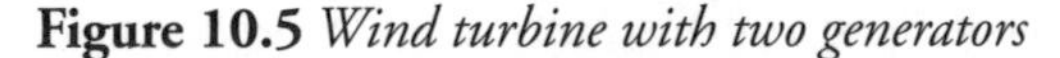
Figure 10.5 *Wind turbine with two generators*

To utilize low wind speeds more efficiently, turbines with two generators were developed, with a small generator for low winds (and a low rotational speed on the rotor) and a large generator with the nominal power of the turbine for higher wind speeds (and higher rotational speed on the rotor).

Source: Typoform

tain good efficiency. The introduction of two generators was the first step towards variable speed turbines.

If a generator is run on variable speed, the frequency of the electric power will also vary. And the electric current has to be adapted to fit the grid. To solve this problem, the generator is 'disconnected' from the grid. The AC from the generator is first rectified to DC and then reconverted to AC by an inverter, which gives the current the same frequency and voltage properties as that of the grid. Wind turbines with variable speed have been available on the market since the late 1980s. They use a synchronous generator combined with power electronics, a frequency converter (see Figure 10.6).

Direct drive generator

Multi-pole synchronous generators, so-called *ring generators*, have been used for a long time in hydropower stations. The advantage of this type of generator is that it can be operated with a low rotational speed. In a wind turbine a ring generator can be driven directly by the rotor without an intermediate gearbox. Such a wind turbine can have a *direct drive* generator.

The German manufacturer Enercon first introduced this design in 1992 on its E-40 model. This ring generator has a large diameter, around four metres, and about 60 poles instead of the four or six that is normal on a standard generator. The ring generator can produce electric power with the same rotational speed as the rotor. Enercon turbines have variable speed and use power electronics to adapt the electric power to the grid (see Figure 10.7).

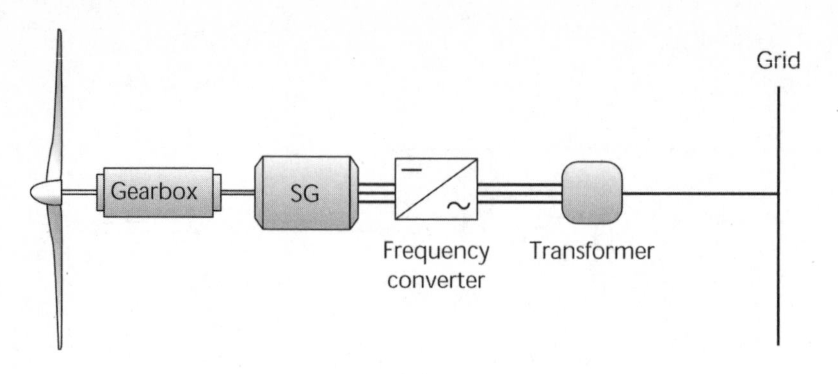

Figure 10.6 *Variable speed turbines*

Wind turbines with variable speed can use a synchronous generator. The AC current is rectified and then inverted back to AC with the same frequency as that of the power grid by a frequency converter.

Source: Typoform (after Tore Wizelius)

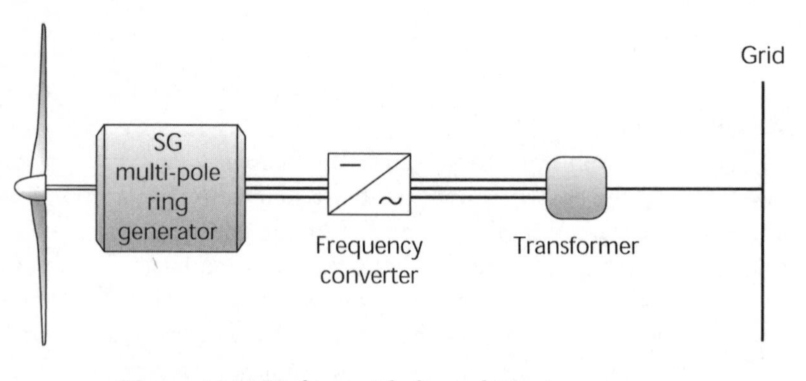

Figure 10.7 *Turbine with direct drive ring generator*

A multi-pole ring generator can be driven directly by the rotor since it has low rpm. Because the synchronous generator operates with variable speed, the current has to be converted to the grid voltage and frequency by power electronic equipment.

Source: Typoform (after Tore Wizelius)

Other manufacturers have also begun to use this design concept, for example the French company Jeumont and the Norwegian ScanWind.

One important advantage with a direct drive generator is that there is no gearbox needing maintenance. The main disadvantage is that large ring generators are very heavy, so the weight of the turbine will increase. An interesting idea that has been applied by the Finnish manufacturer WinWind is to design a hybrid – a wind turbine with a smaller multi-pole ring generator combined with a simple and very robust one step planet gearbox. Several 1MW turbines and some 3MW turbines have been installed and are operating in Finland, Sweden and Estonia.

With variable speed the turbine rotor can utilize the wind more efficiently. The difference in efficiency between turbines with full variable speed and turbines with two fixed speeds is however almost negligible. There are, though, other advantages with variable speed. The wind is always more or less turbulent, so loads on the turbines change all the time, which causes great strain on all components. If the rotor speed is allowed to vary, the power from gusts can be absorbed by the rotor, by increasing the rotational speed (it accelerates) and not be passed on to the main shaft and other components. As turbines get bigger, it will become more important to reduce these loads.

Generator with slip

Manufacturers who use asynchronous generators have begun to design their turbines so that the rotational speed can be allowed to vary within certain limits. The technical solution to achieve this involves either changing the resistance in the rotor windings or controlling the currents in the rotor windings by power electronics. Vestas has used a system named OptiSlip. This makes it possible to increase the 'slip' of the generator rpm from 1 to 10 per cent. When wind gusts give sudden increases of power, the turbine rotor can increase its rotational speed by 10 per cent without affecting the frequency or power output of the generator. The excess power is turned into heat.

The next step in the technical development of the asynchronous generators was to use a slip ring with a rotor cascade coupling (see Figure 10.8). Vestas, Nordex and several other manufacturers now use this system. In an asynchronous

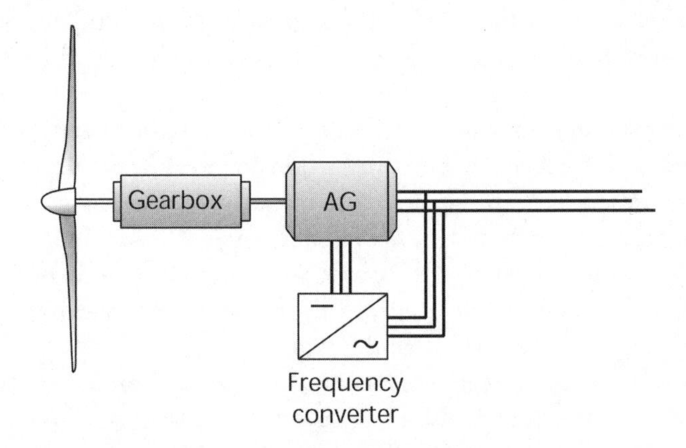

Figure 10.8 *Asynchronous generator with rotor cascade coupling and frequency converter*

Rotor cascade enables variable speed. Most of the power is generated in the stator windings. Only a small share from the rotor windings has to be converted to the grid frequency.

Source: Typoform (after Tore Wizelius)

generator currents are also induced in the rotor. This current has a low frequency, which is governed by the slip of the generator's rotational speed; it is transmitted through slip rings and the frequency converted to the grid frequency. This concept makes it possible to govern the rotational speed over a wide range; in other words it can be variable. In a Nordex 2.5MW turbine the rpm can vary from 11–19rpm. With this solution the frequency converter can be much smaller, since only a small share, about 20 per cent of the nominal power of the turbine, needs to be converted. Another advantage with this system is that the reactive power can be controlled.

Grid connection

The voltage level of large wind turbines is in most cases 690V, so-called industrial voltage. Thus they can be connected to a factory without a transformer. Smaller turbines, up to 300kW, which were common in the early 1990s, have a voltage of 400V and can be connected directly via a feeder cable to a farm or a house. Usually, however, wind turbines are connected to the power grid through a transformer that increases the voltage level from 400 or 690V to the high voltage, normally 10 or 20kV, in the distribution grid.

For small and medium-sized wind turbines a suitable transformer is installed on the ground next to the tower. In large wind turbines the transformer is often a component of the turbine itself. It is often installed in the 'cellar' of the tower, below the level where the door is. Some manufacturers, however, have installed it in the nacelle, where it also acts as a counterweight to the rotor; the cables that transmit the power from the nacelle to the ground, through the tower, can thus be thinner, since high voltage cables need much less cable area than low voltage ones.

The electric system in wind turbines has been developed considerably since the early 1980s; it has been adapted to be 'kinder' to the grid. In modern wind turbines the properties of the electric power that is fed into the grid can be governed by power electronics to achieve the phase angle and reactive power that the grid needs at the point where the turbine is connected. Wind turbines, which once caused power quality problems for the grid operators, can thus now be utilized to improve the power quality in the grid.

However, the power electronics used to solve the old problems have in turn created some new ones (this is also the case with the power electronic equipment used by factories and households). The electronic equipment generates so-called harmonics, currents with frequencies that are multiples of 50Hz (and even some that are not), and this has a negative impact on power quality. This 'dirt' can to some extent be 'cleaned off' by different kinds of filters, but such equipment is expensive and seldom manages to take care of all the 'dirt'.

The control system of wind turbines

Every little movement a wind turbine makes is governed by modern computer technology and 'spikes' and other 'pollution' in the current are screened by power electronic equipment. Turbines are connected via modems to the offices of the owner and manufacturer, who get operational data fed directly to their PCs. If something malfunctions, an operation alarm alerts the owner or operator of the turbines.

The control system fulfils three different functions: operation control, surveillance and operations follow-up (see Box 10.1).

A computer, in most cases installed at ground level inside the tower, governs the control system of a wind turbine. Often there is also a terminal with a display in the nacelle which is used during maintenance. Data collected from anemometer, machine components and the grid are transmitted by fibre optic cables.

The wind is constantly varying – wind direction as well as wind speed will change almost every second – and to achieve efficient production, the turbine rotor should be perpendicular to the wind direction. Thus the control system continuously checks the wind speed and wind direction. This information is processed in the computer, which can instruct the yaw motor to turn the nacelle a

BOX 10.1 CONTROL SYSTEM FUNCTIONS

The control system of a wind turbine has three different functions:

1 **Operation control**
The computer in the wind turbine collects data from the wind vane and anemometer, and if the nacelle has to yaw into wind, it sends a signal to the yaw motor to start working. On turbines with variable pitch blades the control system also manages the blade angle adjustments. It also governs when the generator is connected to or disconnected from the grid.

2 **Surveillance**
Sensors check the temperature in the gearbox, generator and many other components, vibrations in the rotor and nacelle, grid voltage, and many other parameters. When a registered value exceeds its tolerance zone, the turbine will be stopped and an alarm signal sent to the owner/operator by telephone or staff locator. If no serious error has occurred, the operator can restart the turbine from an office PC.

3 **Operations follow-up**
The computer collects data on production, wind speed, outages and many other things. This information is processed and presented in readable form – graphs, tables, etc. – on the display in the turbine or on the office PC.

All medium and large wind turbines on the market have advanced computerized control systems.

specific number of degrees. On turbines with variable pitch, the control system calculates when and how much the blades need to be adjusted.

However, a wind turbine cannot follow all the unpredictable changes of the wind. If it did the nacelle would constantly move to and fro and the yaw motor would wear out very quickly. Operational reliability is a very important property for wind turbines, which are often sited in remote places; turbines should be in operation day and night, all year around, without expensive service and maintenance. Thus the control system has to be designed to optimize not only production but also the useful life of the turbine and to protect the turbine against damage from power outages or when a component breaks or malfunctions.

The control system will therefore not instruct the yaw motor to turn the nacelle until a change of wind direction seems to be lasting. The new wind direction has to change at least a preset number of degrees, and keep that direction for a preset number of seconds or minutes, before the yaw motor is instructed to adjust the nacelle. The control program also has to keep count of how many times the nacelle has turned full circle. After three full revolutions the cables that hang down through the tower will have been twisted into a tight bundle, and the control system then stops the turbine and the yaw motor is instructed to rewind the nacelle and the cables before the turbine is restarted.

The control system keeps close surveillance on all the functions of the turbine. The control system is like a brain, with a nervous system of fibre optic cables with sensors that check the temperature in gearbox and generator, the pressure in hydraulic systems, vibrations in machine components and rotor blades, voltage and frequency in the generator and the grid, and many other parameters. These kinds of data are saved in the computer's memory for a couple of days to enable a thorough analysis after operations disturbances have occurred.

If there is a fault in the functioning of the turbine, the control system will stop the turbine and send an alarm to the operator by telephone or to a staff locator. The operator can then connect his PC to the turbine, check what kind of fault has occurred and print out a report. If the fault isn't very serious, the operator can restart the turbine from his PC.

Wind turbines often have to use the emergency brake even if there is no fault in the turbine itself; a power failure in the grid is the most common cause. Another trivial reason for stops is that a sensor used for surveillance breaks. When a serious fault has occurred, the turbine has to be restarted at the site. The operator has, however, received a fault report to his PC and knows which spare parts he needs to repair the turbine. Often it is a component in the advanced control system that has broken, a printed circuit card or a sensor. The control system itself is one of the most vulnerable parts of modern wind turbines.

Data technology has developed almost as fast as wind power technology, and the cost of advanced computer hardware and software has fallen. By using advanced software to control turbines, efficiency can be increased, and thus software is one of the manufacturer's most valuable assets. Some manufacturers have modem connections to wind turbines installed in different places and can upgrade the software that controls the operations of their turbines from their offices.

The demands on reliability and technical availability will increase even more when wind power is developed offshore. The next step will be to install advanced sensors that can give information about the status of components and give warnings when they start to wear out, so they can be replaced or repaired before they break. Double sets of sensors, so a faulty sensor does not cause operational stops, have already been introduced on many models.

In many countries information about production, fault reports and so on are collected and published in reports or on the internet. To make these kinds of data publicly available has been very valuable for the development of wind power.

11
Efficiency and Performance

How much energy a wind turbine can produce depends on a number of factors: the rotor swept area, the hub height and how efficiently the turbine can convert the kinetic power of the wind. Equally important, of course, is the mean wind speed and the frequency distribution at the site where the wind turbine is installed.

Ever-larger rotors

The power of the wind available to a turbine is $P = \frac{1}{2}\rho A w^3$, in other words power is proportional to the rotor swept area A and the cube of the wind speed w. The rotor swept area on series produced wind turbines has increased at a steady rate and as a consequence so has the rated power of the turbines. Since the early 1980s the power of wind turbines has doubled every 4–5 years on average (see Table 11.1 and Figure 11.1).

Table 11.1 *Development of wind turbine sizes, 1980–2005*

Year	1980	1985	1990	1995	2000	2005
Power (kW)	50	100	250	600	1000	2500
Diameter (m)	15	20	30	40	55	80
Swept area (m^2)	177	314	706	1256	2375	5024
Production* (MWh/year)	90	150	450	1200	2000	5000

* Production on a site with average wind resources.

The rotor swept area has not increased at the same rate as the nominal power of the turbines. The explanation for this is that the towers have also increased in height; a larger rotor needs a higher tower. Since wind speed increases with height, the turbines can catch more power and so it makes sense to use generators with higher ratings. In the 1980s the Danes, after conducting thorough analysis, found that the ideal proportion from an aesthetic point of view between the rotor and the hub height was when the rotor diameter was equal to the hub height.

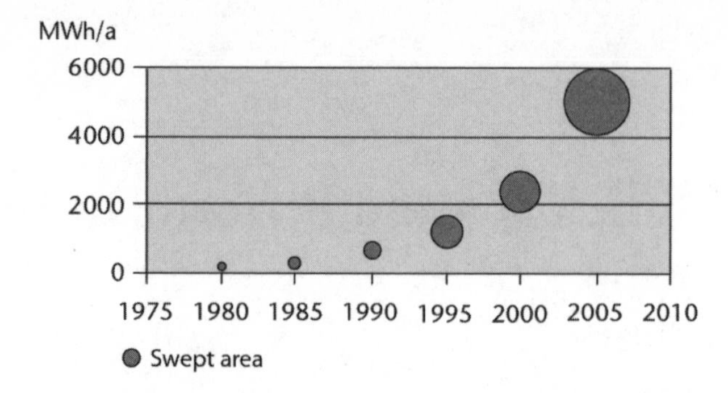

Figure 11.1 *Development of rotor area and annual production*

The rotor area of wind turbines has increased from barely 200m² in 1980 to 5000m² in the first years of the 21st century. Nominal power increased from 50kW to 2.5MW in the same period. Annual production (vertical axis) has increased at a corresponding rate, from about 90MWh/year to 5000MWh/year. In other words, the size of the turbines (the nominal power of the turbines that have dominated the market) has doubled about every 4–5 years. In 2005 there are wind turbines with a nominal power of 4–5MW available; if development continues at the same rate, there will be 8–10MW turbines by 2010.

Source: Tore Wizelius

It is not easy to increase the size of a wind turbine and keep the costs at a competitive level. When the size of wind turbines is increased, these larger turbines have to produce power at a lower price; otherwise it would be pointless to increase the size. The problem is that when the radius of the rotor is increased, the swept area will increase by the square of the radius, while the volume and weight will increase by the cube. The rotor blades and all other components that have to be scaled up all have length, width and thickness. If a small turbine is scaled up to a larger size just by increasing the proportions of all components, the weight will obviously increase much faster that the swept area. And the price is proportional to the weight.

This seemingly irresolvable problem has however been overcome by advanced engineering, more accurate calculations of loads, and development of new design concepts, control strategies and materials. One way of looking at this is that to grow turbines had to go on a diet to reduce the increase in their weight.

Two factors have made this growth possible. First, cost-efficiency has been increased by the increase of power with height. And second, the early turbines of the 1980s were over-dimensioned, which has made it possible to cut weight from most components. In some cases, however, the diet was too strict, which has caused some very comprehensive and expensive (for manufacturers as well as owners) retrofits of the gearboxes of some wind turbine models (see Figure 11.2).

It is impossible for a wind turbine to utilize all the power in the wind. How large a share of the total power in the wind a turbine will utilize is indicated by the so-called *power coefficient*, C_p. The maximum value of C_p is 0.59 (according to

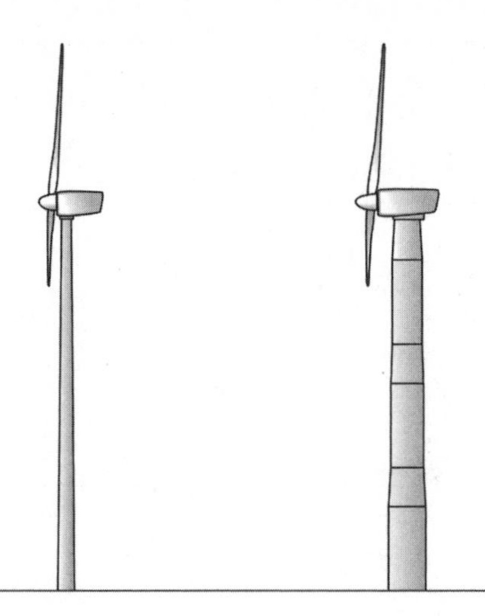

Figure 11.2 *Growth through slimming*

The figure shows a 2MW turbine to the left compared to a 55kW turbine from 1980 that has been scaled up to a comparable size. The 2MW turbine is much slimmer and more compact. This implies that it should be possible to manufacture slimmer and cheaper small turbines today than in the early 1980s.

Source: Stiesdal (2000)

Betz' law, see Chapter 6). The power coefficient varies with the wind speed, and on most turbines the maximum value that can be attained is 0.45–0.50 at a wind speed of 8–10m/s. Most turbines are optimized for these wind speeds, which are often the most frequent wind speeds in terms of the frequency distribution for a year. The tip speed ratio on most turbines is set below the optimum value to limit the swishing noise from the rotor blades.

To convert the power from the revolving rotor to electric power, it passes through a gearbox and a generator, or, for direct drive turbines, through a generator and an inverter. In this conversion some power will be lost to heat. Furthermore, the efficiency of the gearbox, generator and power electronic equipment will vary with the wind speed.

Generator efficiency

A generator is most efficient when it is running at its nominal power. On a wind turbine, most of the time the generator runs on lower power, when the wind speed is lower that the nominal wind speed. The generator is then said to be running on *partial load*. On a standard generator efficiency will then be reduced (see Box 11.1).

To minimize these losses and increase the efficiency of the rotor at low wind speeds (by reducing the revolution speed and using a better tip speed ratio), many

BOX 11.1 GENERATOR EFFICIENCY

Reduction in a generator's efficiency at partial load

% of full load	5	10	20	50	100
Efficiency	0.4	0.8	0.90	0.97	1.00

There is also a relationship between the physical size of a generator and efficiency: efficiency increases with the size of the generator, since losses to heat are reduced.

Relationship between size and efficiency

Nominal power (kW)	5	50	500	1000
Efficiency	0.84	0.89	0.94	0.95

A 1MW turbine running at 20 per cent of its nominal power (200kW) has an efficiency of 0.95 × 0.90 = 85 per cent. Note that the relationships between efficiency, size and partial load can vary between different manufacturers and models; the above figures are typical.

models have two generators with different nominal powers and revolution speeds. Many manufacturers offer one option with just one big generator (a high wind model) and another with double generators (a low wind model), which is more expensive but can utilize the power in low winds more efficiently.

The gearbox

On a large modern wind turbine, the rotor has a rotational speed of 20–30rpm, while the generator will need to rotate at 1515rpm. To increase the speed a gearbox is used. If the turbine rotor runs at 30rpm a gear change of 30:1520 = 1:50.7 will be needed. One revolution of the main shaft has to be increased to 50.7 revolutions on the secondary shaft that is connected to the generator. The gearbox has one fixed gear change ratio (you can't change gears like in a car). With double generators with 1000 and 1500rpm respectively, different speeds from the turbine rotor are used for the two generators. In this case the rotational speed for the small generator for low wind speeds (for the small generator with six poles) will be 20rpm.

A gearbox generally has several steps, so the rotational speed is increased stepwise. Losses can be estimated at 1 per cent per step. In wind turbines three-step gearboxes are usually used and the efficiency of the gearbox will then be around 97 per cent.

Wind turbines with a direct drive generator and variable speed don't need any gearbox. Instead the frequency and voltage of the electric current will vary with the rotational speed. The current therefore has to be rectified to DC (direct current) and then converted by an inverter to alternating current (AC) with the same frequency (50Hz in Europe, 60Hz in the US) and voltage as the grid. The efficiency of such an inverter is also about 97 per cent; losses are therefore about the same as for a turbine with a gearbox.

Overall efficiency

The overall efficiency for a wind turbine is the product of the turbine rotor's power coefficient C_p and the efficiency of the gearbox (or inverter) and generator.

$$\mu_{tot} = C_p \cdot \mu_{gear} \cdot \mu_{generator}$$

Sometimes C_p is set to 0.59 and μ_{rotor} (μ_r) is used to show how large a share of the theoretically available power the rotor can utilize. If the power coefficient C_p = 0.49 the rotor efficiency will be $\mu_r = 0.49/0.59 = 0.83$.

The efficiency of a wind turbine varies with the wind speed. When the wind speed is below the nominal wind speed, the efficiency of the generator will decrease, and if the turbine has a fixed rotational speed, the tip speed ratio will change meaning that the C_p is also reduced. When the wind speed is higher than the nominal wind speed, some of the power in the wind will be spilled: an ever-smaller share of the power in the wind will be utilized and C_p will decrease successively. Wind turbines are used to convert wind to electric power, and therefore another coefficient is used, C_e, which shows how large a share of the power in the wind is converted to electric power at different wind speeds (see Figure 11.3).

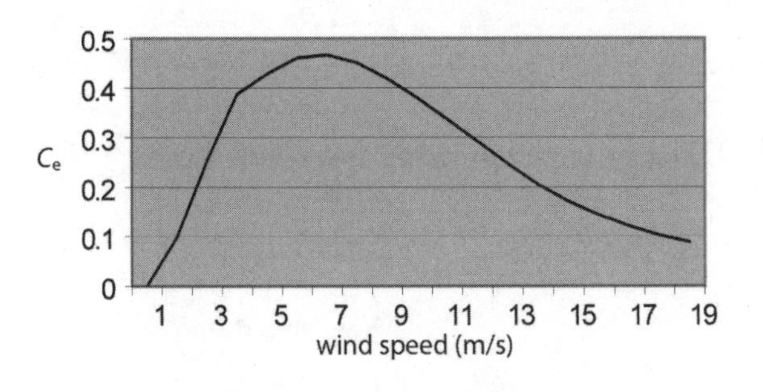

Figure 11.3 *C_e-graph for a Siemens 1300*

A C_e-diagram shows how large a share of the power in the wind is converted to electric power at different wind speeds. The turbine is most efficient at wind speeds between 6 and 8m/s.

Source: EMD (2005)

Power curve

A power curve shows how much electrical power a wind turbine will produce at different wind speeds. The curve can be calculated if the efficiency of the different components at different wind speeds is known. However, the curve also has to be verified by measurements when the turbine is online. There are very specific rules for how such measurements should be performed, and independent certification institutes or companies carry them out to verify the power curve.

The wind speed is measured by an anemometer at hub height on a measurement mast erected at a suitable distance from the turbine and the power from the turbine is measured simultaneously. During the measurement period all wind speeds, from calm to > 25m/s, have to occur for a specified time. The results from these measurements are entered into a diagram, with wind speed on the x-axis and power on the y-axis. Each measurement results in a dot, and together they form something that is far from an even curve, resembling more a swarm of mosquitoes!

The reasons for this are first that there is a short delay before the rotor can catch a gust and turn it into an increase of power and second that when the wind slows down the power is kept at the present level for a short time due to the force of inertia on the revolving rotor – if the wind suddenly calms down completely the rotor will continue to revolve a turn or so before it stops. By calculating the average power for different wind speeds, however, a nice smooth curve can be formed (see Figures 11.4, 11.5 and 11.6).

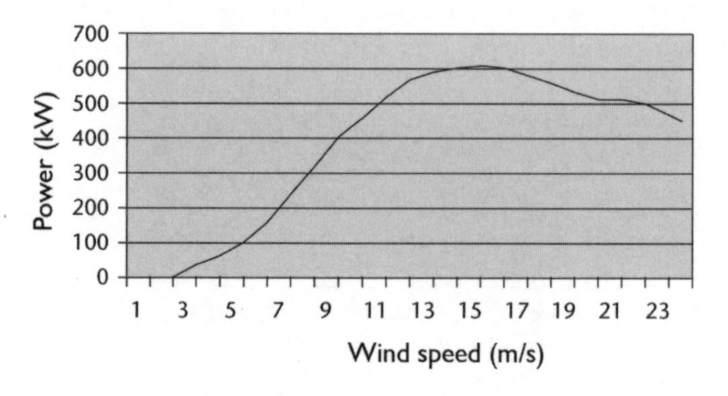

Figure 11.4 *Power curve for a Bonus 600 MkIV*

This 600kW turbine from Bonus (now Siemens) is stall-controlled. The power increases relatively smoothly up to 500kW at 12m/s, where the blades begin to stall. The power continues to increase up to 600kW at 15m/s (nominal power and wind speed) but decreases after that when the stall increases. With fixed rotor blades only the form of the aerofoil controls the power, and it is hard to get a level curve above the nominal wind speed. On earlier models the power dropped more dramatically, but by developing the aerofoils it has been possible to make the curve straighter.

Source: EMD (2005)

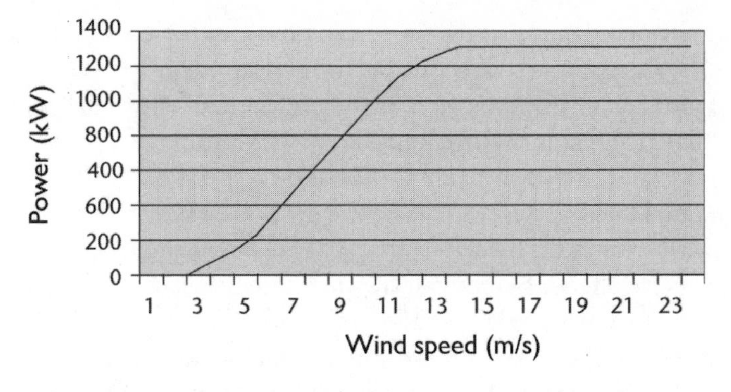

Figure 11.5 *Bonus 1300kW – active stall control*

The Siemens 1300 controls the power by active stall – the rotor blades can be adjusted to control the stall so that the power curve becomes straight above nominal wind speed.

Source: EMD (2005)

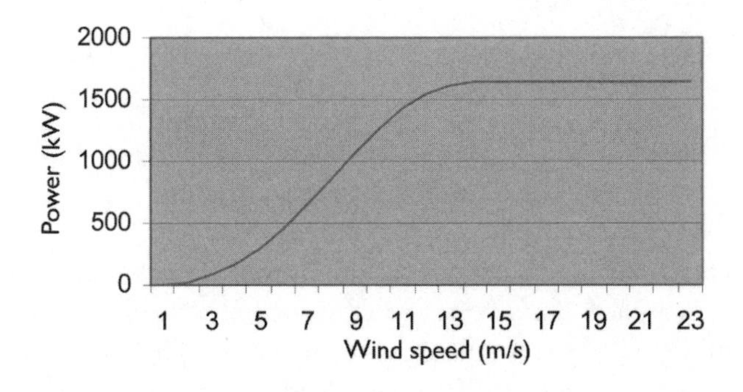

Figure 11.6 *Vestas V66, 1650kW with pitch control*

The Vestas V66, 1650kW is pitch controlled and has a smooth power curve that is straight above the nominal wind speed.

Source: EMD (2005)

Production capacity

To calculate how much a certain wind turbine can be expected to produce at a specific site, it is necessary to know the frequency distribution of the wind speed at that site at the hub height of the turbine. A histogram or table of the frequency distribution shows how many hours a year different wind speeds occur. The power curve for the turbine shows the power it will give at these different wind speeds. Thus the estimated production in a year can be calculated by multiplying the frequency distribution by the power curve (see, for example, Figure 17.2, page 226).

How much a wind turbine will produce depends to a high degree on the wind conditions at the site where the turbine is installed. This makes it quite difficult to compare the efficiency of different turbine models, since they are installed at different sites. Furthermore, it is not only the mean wind speed at the site that matters, the frequency distribution is equally important. And finally it is not the technical efficiency that is decisive, but the cost-efficiency.

It is, however, possible to tailor a turbine for a specific site. There are several options for hub heights. Close to the coast or offshore, a low tower may suffice; on an inland site the tower for the same turbine has to be much higher to get the same production. For a site with a low mean wind speed, it pays to choose a turbine with a large rotor and a low nominal wind speed. For turbines in offshore wind farms the power of the generator and the nominal wind speed can be increased instead. Besides the hub height, turbines can be adapted to a site by choosing a good relationship between the size of the rotor and the size of the generator.

Key figures for efficiency

In wind power statistics different key figures are used to estimate efficiency:

Power production/nominal power, kWh/kW.

Power production/swept area, kWh/m^2.

In both cases mean values for a year are used. However, neither of these key figures on its own gives a good estimate of the efficiency of a turbine. A turbine with a large rotor compared to the nominal power achieves high production in relation to its nominal power, while a turbine with a small turbine and a large generator will produce a lot in relation to its swept area. You can achieve a very high value for one of these key figures by choosing a bad relationship between rotor size and generator size, but a good turbine should offer good values for both key figures.

Capacity factor is a third key figure. This is the mean value of the power of a turbine during a year compared to its nominal power. The same measure is sometimes expressed as *full load hours*. If the capacity factor is 0.3, this corresponds to 0.3 × 8760 hours: 2628 full load hours.

All these key figures say more about the wind resources than about the efficiency of the turbines, however.

The most important key figure is the economic one: total investment/kWh a year. This figure sets the total investment in relation to the production at the site, so it is a useful tool to compare different models, sizes and configurations at a specific site (or several comparable sites). However, since maintenance costs are not accounted for, it does not give a final answer as to what the most economic option will be.

The conversion efficiency for a wind turbine is not so important. The fuel, the wind, is abundant and free. It is the cost-efficiency that matters. It is not reasonable to compare the efficiency of a wind turbine with that of conventional power plants.

Finally there is a key figure for the technical reliability of a wind turbine: *availability*. This figure is given as a percentage. If the wind turbine is out of operation due to faults and ordinary service and maintenance for five days in a year, the technical availability is 98.6 per cent. This means that the turbine could produce power for 98.6 per cent of the time if there was always enough wind to make it run.

The technical availability of wind turbines on line in Sweden is very high: 98–99 per cent. However, component failures do occur when turbines get older. A large share of the disabled time for wind turbines is not caused by faults in the turbine itself, but by unpredicted loss of power in the grid. According to a report from 1997, the most common reason for downtime in Sweden was problems with the gearbox (18 per cent) followed by faults in the control system (15 per cent).

The technical lifetime for a turbine is estimated at 20–25 years. The economic lifetime can be shorter, however, if the costs for maintenance increase too much when the turbine gets old. A turbine at a good site can 'pay back' the energy that has been used in its manufacture in 3–4 months, a very good energy balance compared to other power plants.

The rapid technical development of ever-larger turbines, which have often been introduced on the market without adequate test periods, has caused some expensive mistakes. Many gearboxes on turbines in the 500–1000kW range turned out to be too feeble and have poor lubrication. Thousands of gearboxes had to be retrofitted, an operation that turned out to be very expensive both for the

Table 11.2 *Key figures for wind turbines*

	Calculation	Units
Power/swept area:	$\frac{\text{Production per year}}{\text{Rotor swept area}}$	kWh/m^2
Power/nominal power:	$\frac{\text{Production per year}}{\text{Nominal power}}$	kWh/kW
Capacity factor:	$\frac{\text{Production per year}}{\text{Nominal power} \times 8760}$	%
Full load hours:	$\frac{\text{Production per year}}{\text{Nominal power}}$	hours
Cost-efficiency:	$\frac{\text{Investment cost}}{\text{Production per year}}$	cost/kWh/year
Availability:	$\frac{\text{8760 hours} - \text{stop hours}}{\text{8760 hours}}$	%

manufacturers and for the owners. In the US, where new models are introduced on the market even faster than in Europe and sales departments market models before they have even been manufactured, the largest wind power manufacturer, Kenetech Windpower, went bankrupt when it turned out that their rotor blades cracked – they could not stand the wind.

The quality of the components in wind turbines can vary widely. According to a Swedish technical inspector who examines wind turbines, some components have Rolls-Royce quality, they are virtually indestructible, while others could have come from a discount firm for do-it-yourself customers. All manufacturers have this mix of high and low quality components, but not of the same components, he claimed. There is obviously more to be done when it comes to quality control by the manufacturers and their subcontractors.

However, the technical availability of mass produced wind turbines is very high, and the technical lifetime is estimated at 20–25 years. Furthermore, when a turbine has served its time, it can be dismantled and most of the components can be recycled.

PART IV

Wind Power and Society

To install wind turbines it is necessary to have permission from the local municipality, regional authorities and in some cases also the government. The authorities will assess whether the applications are in accordance with laws and regulations. The legal framework and the economic rules for production, trading and distribution of power are defined by the energy policy of national governments. In Chapter 12, Wind Power Policy, laws and rules for wind power are described.

Wind power is a renewable energy source that will reduce the environmental impact from power production. At the local level, however, wind turbines can give rise to noise and shadow flicker that can have an impact on neighbours. The impact on the environment from wind turbines is described in Chapter 13, Wind Power and the Environment.

To develop wind power in a proper manner, with the energy content in the wind utilized efficiently and conflicts with other interests avoided, spatial planning is necessary. How such spatial planning can be undertaken is described in Chapter 14, Wind Power Planning. Wind power development can also be controversial, and many projects cause a heated debate in the media. In Chapter 15, Opinion and Acceptance, the results of opinion polls and other investigations are described.

Finally in this section, the role of wind power in the power system, the infrastructure for distribution of electricity, is described in Chapter 16, Grid Connection of Wind Turbines.

12
Wind Power Policy

The preconditions for the development of wind power are set by the national policies of a country and by the laws and regulations that are inaugurated by its parliament. And politicians also decide the economic rules of the game: taxes, charges, subsidies and other means of control often affect energy prices, on top of the actual cost of power production. There are also different ways to promote wind power and other renewable energy sources. In short, politicians set the framework for wind power development and thus influence the pace of development.

Each country has it own laws prescribing the procedures and permissions that are necessary to get permission to install wind turbines, and setting the framework for the economic conditions for the sale and distribution of wind-generated electricity. I am familiar with the procedures and the economic framework applied in Sweden and some other European countries, but it is impossible for me to cover the situation in all countries. Therefore here I try to give a general view, with examples mainly from Scandinavia and other parts of Europe, and hope that what I say applies to most other countries as well.

To install a wind turbine it is in most cases mandatory to obtain building permission from the municipality, in some cases from the regional authorities – counties and equivalent – also, or even from the government. The authorities will assess if the applications are compatible with laws and regulations.

All countries have some kind of energy policy which has been formulated by the government and approved by the parliament.

At the international level there are also agreements, directives and treaties concerning energy. In 2001, for example, the EU adopted a new directive on renewable energy with recommended targets for its member countries. The Kyoto Protocol, which was ratified in 2005, also set obligations for industrialized countries to reduce their emissions of greenhouse gases, and development of wind power is considered to be an effective means to achieve that end. These kinds of international treaties have an impact on the energy policy of specific countries.

Permission inquiry

The *municipality* (or council – terms will of course vary from country to country) takes up a position regarding planned wind power projects by evaluating applications for building permission. The building committee or similar institutions usually have this task, and the members of the board of the committee are local politicians that represent the inhabitants of the municipality. The decisions taken have to conform to relevant laws, in this case usually the building law.

The *county administration* in most countries is a state authority at the regional level. For larger projects (in Sweden this means projects of 25MW or more from December 2006), this may be the level at which the decisions on permissions to develop wind power projects have to be taken. The county administration undertakes a legal inquiry to see if the project conforms to the relevant laws.

In other words, all wind power projects will need permission from the municipality; large projects may need an additional permit from the regional or even the national authorities.

Environment Impact Assessment

For larger projects it is compulsory to undertake an *Environment Impact Assessment* (EIA). The rules for the size of a project that demands a comprehensive EIA vary from country to country (see Table 12.1).

The processing of applications to get permission for wind power development differs according the laws and procedures of different countries. In Sweden this process usually takes at least one year, and in many cases several years if there are appeals against the decisions made by the authorities.

Table 12.1 *Project size and EIA-demand in Sweden, Denmark and Germany, 2005*

Country	Local municipality	Regional authority – EIA	Central authority
Sweden†	< 1MW	1–10MW	> 10MW
Denmark	1–3 turbines < 80m hub height	4 or more turbines or > 80m hub height	
Germany	1–2 turbines	3 or more turbines > 10MW*	

* Can be demanded; mandatory for > 20 turbines (see also Box 14.3, page 187).

† In Sweden the rules were changed from 1 December 2006, so that municipalities can decide for projects up to 25 MW.

A building permission from the municipality is always necessary. For larger projects it is necessary to conduct a comprehensive EIA and to get permission from a higher level, a regional or central authority. The rules for when this is necessary differ in these three countries.

Source: Hansen (2005)

Opposing interests

When an application is processed and evaluated, all possible impacts and conflicts with other so-called opposing interests are investigated. The authorities therefore have to refer the application to a number of other authorities for consideration. The most common opposing interests are:

Neighbours

Impact from noise and rotating shadows from wind turbines can be annoying for neighbours if the turbines are installed too close or in an unsuitable direction in relation to dwellings or holiday cottages. The project developer usually makes estimates of these impacts. Rules and methods for calculation of noise and shadow impacts are described in Chapter 13.

The military

Wind turbines can interfere with military installations for radar surveillance, radio communication and so forth. In some areas the air force will have objections to high structures. Since many military installations are secret it is not always a simple matter to know what areas should be avoided for this reason. You can't simply ask the military where their secret installations are sited! The only feasible method here is by trial and error.

The military's opinion carries a lot of weight; indeed in practice it is a veto. In Sweden and some other countries, however, studies to find out if and how much interference wind turbines have on military installations have been conducted, and these studies show that the interference is far less than previously thought.

Telecommunications systems

There is a risk that wind turbines can interfere with signals from radio, TV and telecommunications masts and (civilian) radar. These kinds of interference, however, have also proven to be far less than feared. Interference on television can easily be taken care of by installing a so-called slave station.

The tower of a wind turbine can actually be utilized as a telecommunications mast that will generate additional yearly revenue for the owner. And telecommunications operators don't have any reserved right for using the air space above private grounds. Applications are, however, as a rule sent to the telecommunications authority and to local telecommunications operators for consideration.

Safety

To reduce the risk of accidents if, for example, ice is thrown from the rotor blades, a blade or other parts fall off, or the turbine falls over, a safety distance may be necessary. Usually it is up to local authorities that grant building permission to decide on safety distances to different kinds of installations, residential areas, etc. A recommended safety distance to a larger road, railway or power line might be the total height of the turbine plus a 50-metre extra safety margin. These kinds of incidents are, however, extremely rare. In arctic regions ice can build up on a turbine and the rotor, but this will generally just cause a standstill. When the ice melts and falls off it will most likely drop to the base of the tower. Nevertheless, in areas with extreme weather conditions, hurricanes and severe storms, some wind turbines have been damaged and even fallen over, in Japan and India, for example, and this has even happened in Denmark.

Civil aviation

For civil aviation there are very well defined and strict rules regulating safety. Some of these concern the minimum distance and maximum height of wind turbines and other structures in areas surrounding airports. These rules are available from the national civil aviation authorities.

Protected areas

In most countries areas that are especially valuable for nature, cultural heritage, recreation or some other common purpose should be protected according to environmental laws and similar applicable laws and regulations. These areas are usually defined by national and/or regional authorities, which also have the task of protecting them. There are also areas that are protected by international treaties. For offshore wind power plants the interests of shipping, fisheries and marine nature protection should also be considered.

Assessments

The fact that an area is protected for some specific purpose – for example as a recreation area or a Natura 2000 area (the EU network of sites designated by Member States under the birds directive and under the habitats directive) – does not by definition exclude wind power development within or close to that area. Whether wind turbines can be installed in such an area will depend on the actual impact of the turbines on the assets or values that are under protection therein. All protected areas have a purpose. If wind turbines don't have a negative impact on this specific purpose, there are no legal grounds for excluding them from the area. These kinds of assessments are made and evaluated in the EIA for specific projects.

Permission process

The permission processes in different countries are quite similar, at least in Europe. However, in some countries the time from application to permission is relatively short, while in other countries it can take years. Another difference is the likelihood of a positive outcome, which also differs greatly. Even if the demands are similar, building permission and an EIA that has to be approved, the efficiency of the process itself differs.

The lead time for a wind turbine installation is only three months. If all processes moved on according to schedule, it should not take more than six months from application until the wind turbine is online. However, the time it takes to realize a project is also a political matter; it is a matter of priority and planning.

In Denmark, where development on land was very fast during the 1990s, the government commissioned the municipalities and counties to find suitable areas for wind turbines within their regions. During this planning process much of the assessments with regard to conflicting interests were made. By this *positive* planning, there were only certain practical details left for discussion when applications were submitted. And the Danes have applied a similar method for their offshore developments, by creating wind fields at sea for this purpose.

In Sweden and most other countries such planning has not taken place. It is up to the developer to find a site and then hope that the authorities agree that it is suitable. Central authorities like the Swedish Environment Protection Agency, the National Heritage Board and the National Board of Housing, Building and Planning, have mapped out areas were wind turbines should *not* be built, in other words have undertaken *negative* planning. So there are no guidelines for where the chances to get the necessary permissions are good. For each and every new project a very comprehensive EIA has to be made.

In the UK the situation is similar to that in Sweden. However, for offshore developments the UK government found a brilliant solution: it created a *one-stop shop*.

An office was established where all aspects and permissions were processed in close dialogue with the developer. The offshore wind farm North Hoyle got permission to start project preparations in April 2001, in February 2002 building and environment permissions were approved and the wind farm was online in December 2003. The share of approved applications increased from 56.5 per cent in 2000 to 96.1 per cent in 2003. In Spain a similar method has been used at the regional level. It seems that to reduce the time for the application process is a matter of political will and administrative skill.

Wind power politics

In Europe the development of commercial wind power started in the early 1980s. Today countries like Denmark, Germany and Spain have several thousands of MW of wind power online as well as a new industry with thousands of employees. In other countries, like the Netherlands, the UK and Sweden, development has been considerably slower. It is not very difficult to find explanations for the large differences in the pace of development between different countries. The explanation can be found in the wind power policy that has been conducted by different governments.

Today it is perfectly clear what kinds of political measures, rules and regulations promote rapid development of wind power and what measures can be used to keep development at a slow and low level.

In Denmark, Germany and Spain it has been profitable to invest in, to own and to operate wind turbines. The laws and regulations in these countries guarantee a fixed and relatively high price for the power produced during the time period that it takes for the owners to get their money back. There have been clear political signals that wind power should be developed fast, and it has not been too difficult and time consuming to get the permissions necessary to install turbines and connect them to the power grid.

Politicians in these countries have had several different motives for their support for wind power. Renewable energy from wind power will reduce the direct impact power production has on the environment, as well as the emissions of carbon dioxide that most countries have been obliged to reduce in accordance with the Kyoto Protocol. Wind power will thus enable politicians to attain these goals (see Box 12.1).

BOX 12.1 EU TARGETS FOR RENEWABLE ENERGY

The EU has been working towards the general target of an increase to 12 per cent of renewable energy's share of gross inland consumption in the EU by 2010 (compared to 6 per cent in 2001). To provide a focus for faster progress, the EU has adopted an operational target for renewable energy in the electricity sector through its Directive 2001/77/EC on the Promotion of Electricity Produced from Renewable Energy Sources in the Internal Market. The directive aims at increasing the share of electricity generated by renewable energy for the EU (including wind power) to 22 per cent in 2010 from a level of 14 per cent in 2000.

Source: EC (2001)

Comprehensive view

An equally important factor, probably, is that wind power has been considered as a new and very promising industry with a huge growth potential that can give a considerable pay-back to the economy by increasing employment, economic growth and export revenues. In Germany the support to wind power has also been a part of their agricultural policy. Wind turbines give an additional income to farmers, meaning that they can make a living and stay in the countryside instead of having to move to town and line up among the unemployed.

Even if the economic subsidies for wind power during its early stage of development are relatively expensive for the economy, politicians have calculated that in the longer run it will generate economic benefits. In countries where the demand for electric power is increasing, this is sufficient reason for developing wind power, since it is one of the cheapest ways to produce electric power today, given sites with good wind conditions.

This comprehensive view, taking into account industrial policy and national economics, has been lacking in the Netherlands, the UK and Sweden. In these countries the domestic wind turbine manufacturers languish due to the lack of a growing and stable domestic market. The rules for power sale and the permission processes have been complicated, and the purchase price for wind-generated electric power has been low. One of the aims of such a policy has been to reduce the cost of wind-generated power to the lowest possible level, with the laudable ambition to strengthen the competitiveness of wind power. Instead the development pace has been retarded and the domestic manufacturing companies have been nipped in the bud.

New growing industries need good economic margins, since they have to invest large amounts in research and development to be able to keep their competitive strength on the market. Two researchers from Chalmers Technical University in Gothenburg, Professor Staffan Jacobsson and Anna Johnson, in their article 'The development of a growth industry – The wind turbine industry in Germany, Holland and Sweden' (Jacobsson and Johnson, 2003), note that Germany has succeeded in developing a wind turbine manufacturing industry, while the Netherlands and Sweden have failed. They point out four explanations for the German success: diversity, legitimacy, markets and industrial policy.

In the 1980s Germany, as well as the Netherlands, went in for diversity, and within their R&D programmes many different technological concepts were tested, while Sweden put all its money into the development of very large turbines in the multi-MW class.

Both in Germany and the Netherlands wind power technology was considered as a legitimate investment. There was a political consensus on the necessity to develop wind power, and it became legitimate for private capital to invest in this new branch. In Sweden this legitimacy was missing, due to the hostile debate on nuclear power that has lasted for three decades. In this debate renewable energy has been reduced to, and defined as, a substitute for nuclear power. Wind power development has thus been seen as a threat by Swedish industry, especially basic industry (steel and paper mills, etc.), with their huge demands on cheap electricity, but also the power companies that own the nuclear power plants, Jacobsson and Johnsson argue.

In 1989 Germany introduced an ambitious programme to develop 250MW of wind power, which created a domestic market for the German companies that manufactured wind turbines. After a year a new law was passed that guaranteed wind turbine operators high revenue for an extended period for the power that they fed into the grid. This created a fast growth that was perceived as a threat by the German power companies, who started to act to change these new laws.

However, the fast growth of wind power also created a new group of power plant owners: farmers, cooperative associations and other private investors. And at the same time as this new industry was developing, wind power was also gaining political strength and influence. There was strong support for wind power in the German Federal Diet (der Bundestag) that was able to stop the efforts of the power industry lobby to impair the conditions for wind power. The Netherlands also launched a similar 250MW programme, but it failed due to the difficulties of getting building permissions from the municipalities and because the government didn't take any resolute actions to eliminate this bottleneck.

German industrial policy created a quasi-protected market, since the support that was given within the framework of the 250MW programme was distributed to different technical concepts, to evaluate which was best. By this measure 60 per cent of the turbines that received support from the programme could be delivered by German manufacturers, in spite of the fact that the Danish wind power industry at that time had a large lead.

In the Netherlands too the domestic manufacturers were favoured: 90 per cent of turbines were 'made in the Netherlands'. By this exaggerated protectionism, however, the domestic manufacturers were locked in on their domestic market. Sweden has not had any similar ambitions for industrial policy as part of its energy policy, which has been governed by the demand of basic industry for cheap electric power. The market that was created by the investment subsidies introduced in the 1990s was taken care of by Danish manufacturers.

National energy policies

The energy policy of a country governs the conditions for wind power and its ability to compete on the electric power market. If the energy policy of Denmark, where wind power has been a success story, and Sweden, where development still has not taken off, are compared, the differences are easy to identify. In the paper 'Possibility of wind power: Comparison of Sweden and Denmark' (Miyamoto,

2000) the differences in the energy policy of these two neighbouring Scandinavian countries are described and analysed.

In Denmark, which has a partly decentralized power system, the government has had a strong and explicit ambition to develop wind power. A large share of the electric power in Denmark is produced by coal- and gas-fired power plants. In the 1980s generous investment subsidies were available to farmers and other private investors in wind power, who could also get favourable loans since the financial sector did not see any political risk in such credits. Counties and municipalities were given the task of creating physical space for wind turbines by working out local and regional wind power plans, which also had a positive impact on investments.

At the same time the government ordered the utilities to take an active part in the development of wind power and to pay a good price for wind-generated electric power from independent power producers. A very strong domestic market was developed for the wind turbine manufacturers, who also were helped by the government to establish themselves on foreign markets.

Denmark has conducted a long-term wind power policy where economic means of control, planning and other measures have been coordinated. By this policy Denmark has managed to adapt its power production to environmental demands, and the obligation to the Kyoto Protocol, to reduce emissions of greenhouse gases, which was not on the agenda when the development started. At the same time Denmark managed to develop a new industry that today is dominating the world market (see Figure 12.1).

In Sweden, which has a strongly centralized power system, the government has not had any high ambitions to develop wind power. Since a very large share of the power is produced by nuclear and hydro-power plants, there has not been any pressure on the government to reduce emissions of greenhouse gases from power production. The financial sector has not perceived any signals to invest money in wind power and has had a negative attitude to such investment. The rules and regulations for permission to develop wind power are more complicated than in Denmark, which, combined with the Swedish tradition to protect nature and the lack of coordination between authorities at different levels, has made the permit process very drawn-out.

The power industry has been passive and the prices for wind-generated power have been very unstable and unpredictable. Even in Sweden private owners have, with the help of independent project developers, installed wind turbines, and with some subsidies from the state; however, this development has been too modest to create a strong domestic market or a wind turbine manufacturing industry (see Figure 12.2).

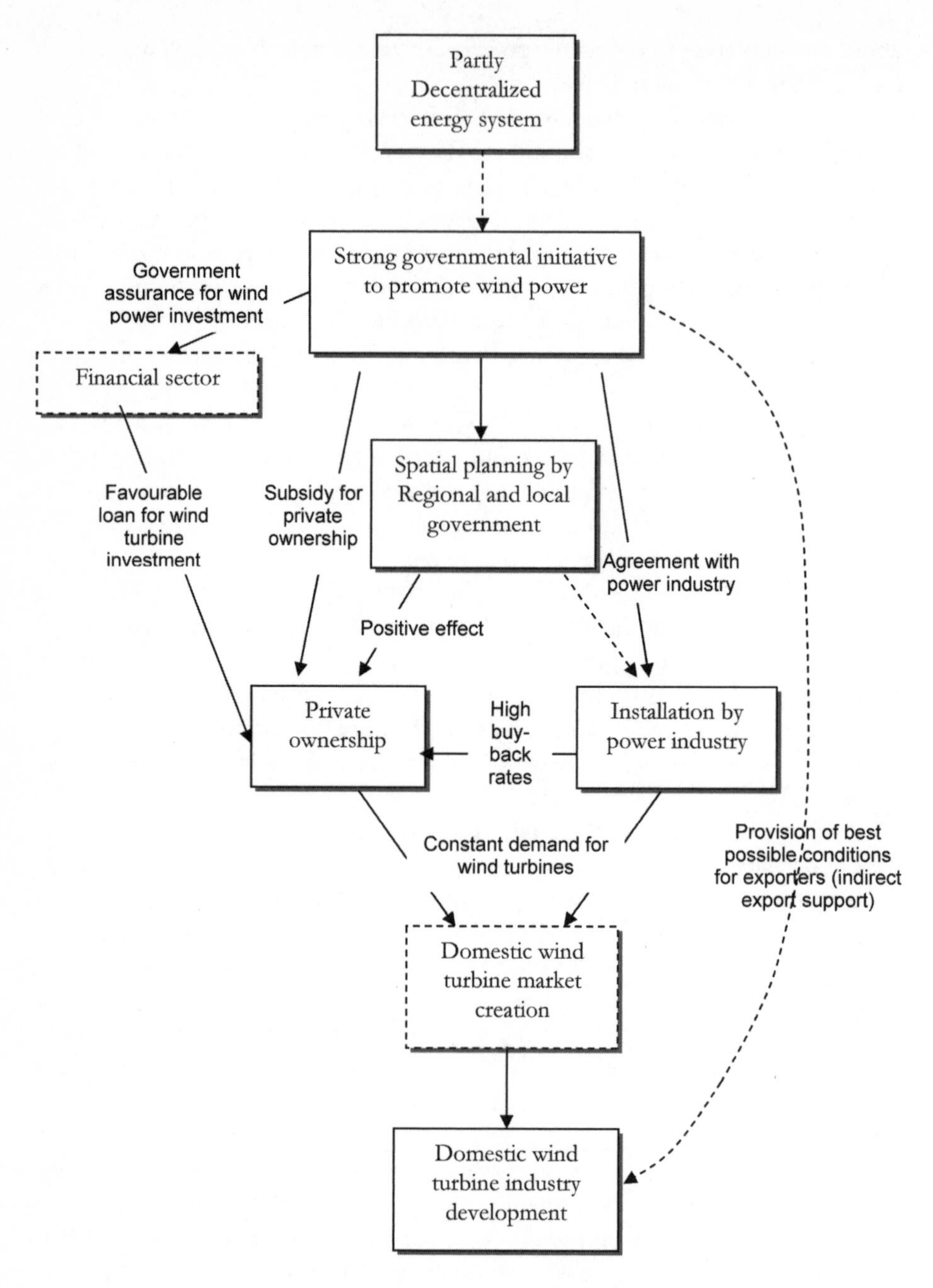

Figure 12.1 *Denmark's wind power policy*

Source: Miyamoto (2000)

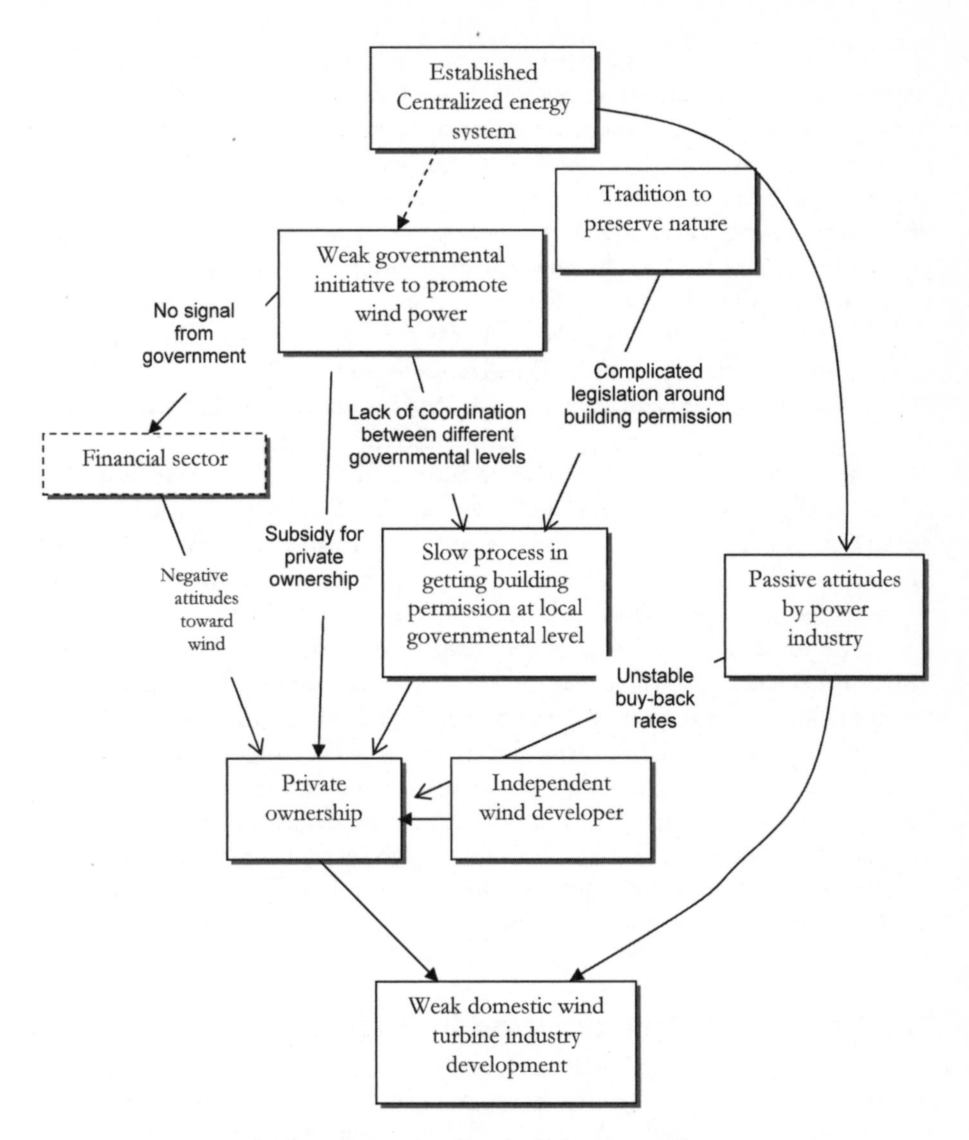

Figure 12.2 *Sweden's wind power policy*

Source: Miyamoto (2000)

Wind power on the power market

In the second half of the 1990s the deregulation of the electric power markets was put on the political agenda in the EU, the US and elsewhere. At that time the supply and distribution of power was a strictly national matter, regulated by the state and considered as a matter of public service and national security for the country.

Deregulation advocates argued for a radical change of approach. The first step would be to separate power production, distribution and trade. Power plants would compete and get an incentive to cut costs, which would push down prices for power. All power producers would get free access to the grid. Finally electric power should be a commodity, bought from power producers and sold to customers by independent power trading companies. The final step would be to open the boundaries, to interconnect the power grids of different countries and regions to create an international power market.

This process towards deregulation has begun, and in some countries, like Sweden, Norway, the UK, and some federal states in the US, like California, the markets have been completely deregulated. Other countries have just started with the deregulation and there are still countries resisting the process.

One important question that has to be solved in the deregulation process is who will have the responsibility for the overall system balance and how this can be maintained. In a power system, supply and demand has to balance at all times, otherwise power cuts would be commonplace. The power crisis in California after deregulation, where the power prices skyrocketed and many huge power companies, like Pacific Gas and Electric, went bankrupt, proved that deregulation is a tricky business. To avoid such situations, someone has to have the overall responsibility and the legal power to order power producers to supply the power needed to balance demand.

There is no doubt that the deregulation process will continue, and that it will work when rules and regulations have been adjusted. Whether this will lead to lower power prices for customers, however, is not self-evident. In a public service system the actual cost of production sets the price, while in a market system it is determined by supply and demand. When demand increases, prices will rise, irrespective of the cost of producing the power.

Deregulation in Sweden

In Sweden a new electric power law came into force on 1 January 1996. The power market was deregulated, or re-regulated, as some prefer to say, since there still is a legal framework that regulates the market. With this new law the monopolies on the sale of power were abolished and power trade was opened to free competition. Every customer can now freely choose from which power company to buy electric power.

Production, trade and distribution of power are strictly separated. These tasks used to be managed by the same company – municipal utilities, the state-owned Vattenfall or private power companies – which had reserved rights to sell and distribute power within their regions of operation. According to the new law, trade in power shall be handled by power trading companies and the distribution of power by grid operators. Many power companies and utilities have created separate affiliated companies to handle the power trade.

Grid operation is still a monopoly, since it is considered too expensive and hardly rational to install competing power grids. The Swedish energy agency issues licences to grid operators, which get sole rights to distribute power in a specific region. Prices for this service have to be reasonable and are supervised by the energy agency. The price for power is, however, completely unregulated, determined only by the market.

A customer of electric power always has to make two separate contracts, one with a power trading company for the purchase of power and another with the grid operator that distributes the power to the customer. The customer pays one fee to the grid operator for the distribution of the power and one to the power trading company for the power. On top of this there are taxes, including electric power tax and finally value added tax. These three shares, grid, power and tax, are of about equal size. The only cost that can be changed by choosing another power trading company is the price for the power, which constitutes a third of the power bill.

The justification for the deregulation of the power market was to create competition between different power producers and power traders so that power prices would be lower. And indeed deregulation led to significantly lower power prices, for a few years. During the same period, however, in the wake of deregulation, the large power companies bought and incorporated most of the smaller companies, so that by the beginning of the 21st century ownership in Sweden was concentrated in three large companies. Since then power prices have increased and reached the same or even higher levels than before deregulation.

In Sweden today the market mechanism is the predominant politico-economic doctrine. Wind power has to compete with other power plants on a free power market. It is, however, always more expensive to produce electric power in new power plants than in plants that have been online for many years. The reason for this is not that new power plants are more expensive, but that older power plants (that often got significant economic subsidies when they were built) have already paid back a large share of their loans and consequently have much lower capital costs. This situation, which makes it necessary to give extra economic support to new power plants, does not apply only to wind power, but to all new power plants.

To make it possible to build new power plants that can replace the old ones when they are taken out of operation, some kind of economic support is necessary.

Support Schemes for Renewable Energy

In the project RE-XPANSION researchers from several European countries have described and evaluated the different support schemes that have been used to promote wind power and other renewable energy sources. The result of their comparative analysis is presented in the report *Support Schemes for Renewable Energy* (EWEA, 2005a). Incentives to promote wind power can be grouped into three categories:

1. **green marketing**: voluntary systems where the market determines the price and the quantity of renewable energy;
2. **fixed prices**: systems where government dictates the electricity prices paid to the producer and lets the market determine the quantity; and
3. **quotas**: systems where the government dictates the quantity of renewable electricity and leaves it to the market to determine the price.

The first of these, green marketing, has proved inefficient. In surveys quite a large share of consumers claim that they would pay a little bit more for power from renewable energy sources like wind power, but when they get this opportunity less than 1 per cent actually choose this option.

Systems with fixed prices or quotas are regulated by law, and are thus compulsory, which makes them more efficient. These can then be divided further, and the report defines five different types of support schemes for renewable energy:

1. investment subsidies;
2. fixed feed-in tariffs;
3. fixed premium systems;
4. tendering systems; and
5. tradable green certificate systems.

Investment subsidies

Investment subsidies have been used by many countries in the early stages of wind power development, for example Denmark, Germany and Sweden, and have proved to be quite efficient. The advantages are that it is a simple system and that the subsidies are paid up front. The support cannot be reduced or withdrawn during the lifetime of the project and this gives security to the investor. The drawback is that it does not differentiate good projects from bad. Investment subsidies have now been abandoned in most countries, but are still used at the regional level in Spain.

Feed-in tariffs

Fixed feed-in tariffs have so far proved the most efficient way to promote wind power. The price paid for wind-generated electric power is fixed, either to a specific value or in relation to the consumer price. The price can be fixed for the lifetime of the turbine or until a specified target is reached. The system guarantees that the investors will always get their money back. It is used by the three countries that have been most successful when it comes to wind power development so far – Denmark, Germany and Spain.

Fixed premiums

Fixed premium systems have been used in combination with other promotion strategies. The premiums, in Sweden called environment bonuses, reward operators for the health and environment costs avoided by wind power. In practice the value of the premium is set in relation to the power price to make wind power competitive.

Tenders

Tendering systems have been used in the UK, in the NFFO (Non-Fossil Fuel Obligation) system. This didn't work out too well, however, since many of the projects that won the tenders were never built, and has now been replaced by a green certificate system. Tendering is still used for offshore developments in Denmark and the UK.

Certificates

Tradable green certificate systems for renewable energy have been introduced during recent years in Italy (2002), the UK (2002), Belgium (2002) and Sweden (2003). Producers get certificates based on the power produced during a year; these certificates can then be traded on a certificate market where the price is set by supply and demand.

Evaluation

In the 'Support Schemes for Renewable Energy' report these different support schemes have been evaluated based on a survey among more than 500 experts from the energy field. According to this survey the most important properties of support systems are *investor confidence* and *effectiveness*. The five support schemes described above have been specified in two versions, one generic and one advanced, where rules that would make the systems more efficient have been added.

In the survey the feed-in tariff system got the top score, with green certificate and tendering at the bottom. Both the generic and the advanced version of the green certificate system got low scores for the most important criteria, investor confidence and effectiveness, compared to other support schemes. The main result of the survey is that the feed-in tariff system is the support scheme preferred by the respondents.

The rules and regulations for these different systems vary; each country actually has its own unique system. This makes it difficult to compare and evaluate the efficiency of the different systems. From a historical point of view it seems obvious that the fixed price system is most efficient, since it has made development in Denmark, Germany and Spain so successful. However, there are other countries that have used the same system with less success, like Greece and France.

The support system is but one of several factors that has to be right. According to the authors of the report, there are four main ingredients in a potentially effective overall promotion strategy for renewables:

1 well-designed payment mechanisms;
2 grid access and strategic development of grids;
3 appropriate administrative procedures and streamlined application processes; and
4 public acceptance.

The systems with tradable green certificates have only been used for a few years, so it may be premature to evaluate their efficiency, since the mechanisms that haven't worked so well may be improved. There have been quite strong arguments for a change from the fixed price system to the tradable green certificate system after deregulation of the market – the quota system is claimed to be more market oriented than the fixed price system. The authors of this report have a different opinion, however:

> *A system where the government fixes quantity and leaves it to the market to determine the price is unlikely to be more 'market oriented' than a system where the government fixes the price and leaves it to the market to determine the quantity.*
>
> *The main difference between quota based systems and price based systems is that the former introduces competition between the electricity producers (e.g. wind turbine operators). Competition between manufacturers of plant (e.g. wind turbines), which is crucial in order to bring down production costs, is present if government dictates either prices or quantities.*

It is possible to achieve a specific target (in MW of new power) by a specific time by giving a guaranteed price for new power produced that will stimulate the necessary investments. This is exactly what Denmark has done with wind power. In Denmark development of wind power actually has reached the goals that have been set several years ahead of schedule. The target set for 2005 was already reached in 1999.

The Swedish certificate system

In Sweden politicians have chosen other means to stimulate the development of renewable energy: green certificates (in Sweden called *elcertifikat* – electric certificates). With the certificate system different kinds of renewable energy sources – wind power, biomass and hydro – will compete with each other, and the market will regulate the price of certificates (see Box 12.2).

Box 12.2 THE SWEDISH CERTIFICATE SYSTEM

Power plants that produce renewable electric power have to be approved and registered by the Energy Agency.

The following renewable energy sources have the right to get certificates:

- all wind turbines (irrespective of size);
- combined heat and power (CHP) plants using biomass, including peat, as fuel;
- small-scale hydropower plants (< 1500kW);
- new hydropower plants; and
- retrofits of existing hydropower plants (only for the increased output).

These producers of renewable electric power, when they have been approved and registered, will get one certificate for each MWh/year (1000kWh/year) from the state.

The Swedish national grid company Svenska Kraftnät controls the production of the power plants.

Source: Swedish Energy Agency (2004)

All power trading companies are obliged by law to have a specific share, or *quota*, of renewable energy in the power they sell to customers. They can obtain this share by buying certificates from power producers and the cost for these certificates can then be passed on to the customers. The sizes of the quotas are decided by parliament and to increase demand for renewable energy the quotas are increased from year to year (see Table 12.2).

Table 12.2 *Certificate quotas in Sweden*

Year	2005	2006	2007	2008	2009	2010
Quota %	10.4	12.6	14.1	15.3	16.0	16.9

Quotas were set only to 2010, which made it impossible to estimate the value of certificates after that. In 2006 it was decided to prolong the certificate system up to 2030. These quotas can be revised and changed by parliament at any time, so there is a high political risk connected to the certificate system.

Certificate prices

There is a guaranteed minimum price for certificates and a penalty price for power trading companies that do not fulfil their quota obligations. Certificates that a power producer is not able to sell, the state will buy back for the minimum price, but only up to 2008. Certificates are issued on a yearly basis, but don't have to be traded in any specific year. There are plans to extend this system to incorporate Norway in a common certificate market. The EU has also clearly expressed an intention to create an international market for the trading of certificates in the future.

The justification for introducing the certificate system was to replace direct subsidies financed through the state budget (in other words by taxes) with a market based system that is paid for by customers. However, the value of the certificates is set and governed by the politicians: the size of the quotas (demand) and maximum price (penalty) are decided by parliament. The same system is actually used in Sweden to regulate the production of milk, where the purpose is to *limit* the amount of milk produced. In other words the certificate system can, if it works, be used either to stimulate or to limit the development of renewable energy and wind power.

Independent power producers

In most countries wind power has not been developed by the traditional power companies but by farmers, economic associations and small limited companies formed to install and operate one or small groups of turbines, so called independent power producers (IPPs). Only during the last few years have some new very large actors entered the stage, mainly to invest in large offshore projects. Even in Denmark and Germany, with a very large share of wind power in the power system, most of the turbines are owned and operated by IPPs. In Denmark there are around 70,000 different owners of wind turbines, many of them private households that are members of wind power cooperatives and farmers (see Figures 12.3 and 12.4).

In Germany and Denmark the traditional power industry (power companies and grid operators) have been opponents of wind power and still are to a large extent. Only during the last few years have some of these big corporations changed their attitude. The German company Eon, which also operates in the southern part of Sweden, has invested large amounts in the Danish offshore wind farm Nysted. And the Danish company Elsam, which owns and operates the other large offshore wind farm in Denmark, Horns Rev, changed their mind about wind power by the end of the 1990s. They realized the business opportunities their experience of wind power held and started their own wind power department that is now developing large offshore wind farms in the UK and other parts of the world.

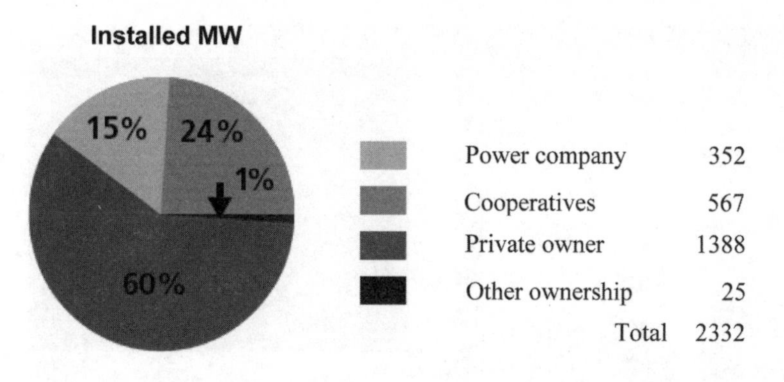

Figure 12.3 *Wind power plant operator categories in Denmark*

In Denmark 85 per cent of the installed wind power capacity is owned and operated by IPPs.

Source: Danish Wind Turbine Owners' Association (2002)

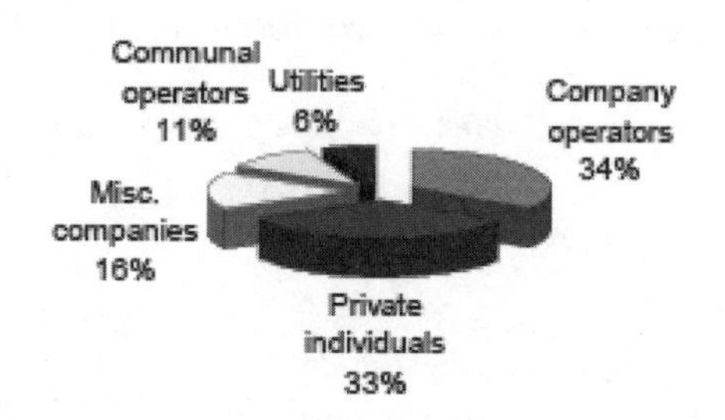

Figure 12.4 *Wind power plant operator categories in Germany*

The shares are based on a sample of plants that were subsidized under the '250MW Wind' programme. These have been quite stable in recent years. In the early development stages, private individuals (mainly farmers) and communal, local citizen-owned operators (in Germany called 'Bürgerwindparks') strongly dominated the wind power market. More recently, increased separation of project development and fundraising, the more complex and demanding nature of larger projects, and the demand for shares in wind park projects from citizens from outside the regions concerned have called for increasingly professionalized company management and funding.

Source: Enzensberger et al (2003)

Competition on equal terms

When the cost to produce wind-generated electric power is compared with the cost of power from other energy sources, the factors that are used for the analysis should be equal; external costs should also be included, since damage to health and the environment by power production generate costs for society.

By the beginning of the 2000s wind power had become competitive, even if the external costs of other power plants were excluded. And its competitiveness will increase even further in the coming years. In reports from 2001 with prognoses for 2020, the UK government and the US Department of Energy have

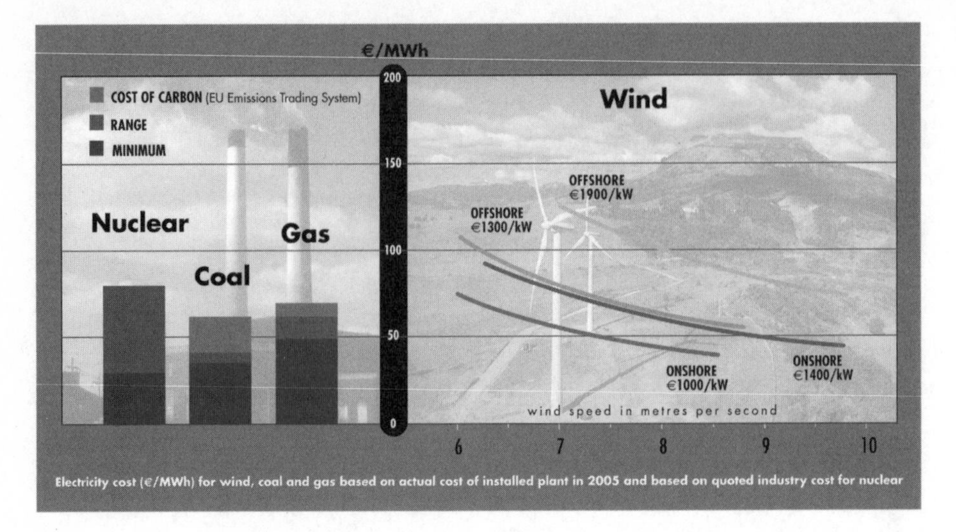

Figure 12.5 *Competitiveness of wind power*

Electricity cost (€/MWh) for wind, coal and gas based on actual cost of installed plant in 2005 and quoted industry cost for nuclear.

Source: Windpower Monthly (2006)

pointed to wind power as the cheapest energy source. In 2005 a wind power station typically cost €1200/kW installed power on land and €1600/kW offshore. The cost of a megawatt hour of electricity generated on land ranged from about €45/MWh at high wind speed sites to up to €80/MWh on sites with a minimal wind resource (see Figure 12.5).

Even today the power prices do not include the external costs. In 2001 the Commission of Europe updated calculations of the external costs of different energy sources in the report 'ExternE: Externalities of Energy' (ExternE, 2002). One of the conclusions was that the costs to produce electric power with coal or oil would double, or increase by 30 per cent in the case of natural gas, if the external costs for impacts on health and environment were included in the power prices (see Figure 12.6).

If external costs are included in the economic calculation, wind power (with the conditions described above) was already the cheapest means to produce electric power with new power plants by the early 2000s. Since the energy source, the wind, is free, there is no risk that the prices will increase due to changes in the world market prices for fuels, which can have a large impact on the costs for power generated by fossil fuels or nuclear power.

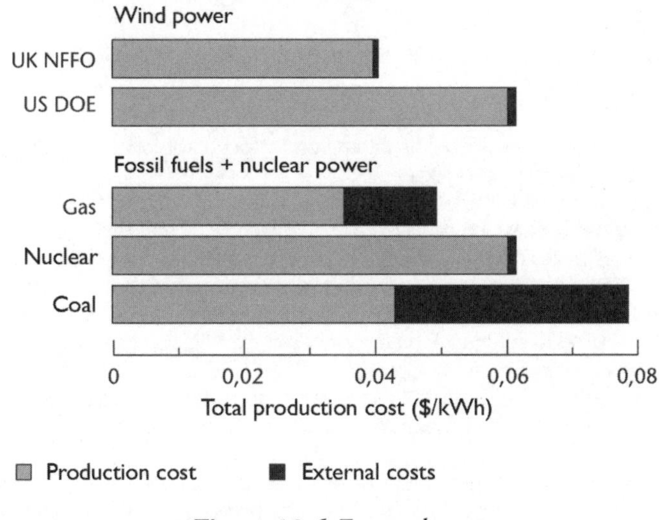

Figure 12.6 *External costs*

The diagram shows the total costs (production cost plus external costs), for different energy sources. For wind power there are two columns, the upper (UK NFFO) shows the actual production cost for wind farms that were developed in England in the NFFO programme, the lower shows the price for wind power in the US, estimated by the US Department of Energy. The external costs are average values from the EU ExternE report.

Source: Milborrow (2002)

Policy recommendations

Wind power is a relatively new power source that differs from conventional technologies in several respects. Wind power plants are favourable for health and the environment, since they don't produce any hazardous emissions. Wind turbines are connected to the distribution grid, and can be used for distributed generation. They are modular, consisting of small units, compared with conventional power plants, and can be added to the power system continuously. Finally they are owned and operated by IPPs.

To be able to compete on equal terms on the electricity power markets, that have been created for conventional power plants and are dominated by traditional, often very powerful, power companies, however, it is necessary with political support to eliminate the competitive disadvantage that has been created by decades of financial and political support for conventional technologies. What kinds of reforms are necessary to accomplish this has been analysed and from the results policy recommendations have been formulated (see Box 12.3).

Box 12.3 Policy recommendations from RE-XPANSION

The *Support Schemes for Renewable Energy* report (EWEA, 2005a) formulates a number of policy recommendations:

- Known external costs should be internalized through appropriate pollution taxes on the polluting energy technologies.
- The costs of climate change are impossible to quantify, so this should be dealt with in a quantity regime which can assure adherence to safe maximum concentrations of greenhouse gases (GHGs) in the atmosphere (by international treaties like the Kyoto Protocol, etc.).
- If external cost cannot be internalized by emission taxes, feed-in tariffs or premiums should be used to internalize the differences in external costs between conventional and renewable energy technologies.
- While reduction targets for GHGs are far from securing safe maximum GHG concentrations, additional measures should be taken to level the competitive playing field for renewables.
- Other reasons for the promotion of renewables – security of supply, diversity of supply, local employment, etc. – should never be overlooked. Such factors need to be taken into account in the making of renewable energy policies in their own right.

The 'Wind Force 12' report, written by Greenpeace for the Global Wind Energy Council, calls for 'legally binding targets for renewable energy'. Such targets will force governments to develop financial frameworks, grid access regulation, planning and administrative procedures. They also need to be accompanied by policies that eliminate market barriers and attract investment capital. The market has to be clearly defined in national laws and include stable, long-term fiscal measures which minimize investor risk and ensure an adequate return on investment.

Further, the report calls for electricity market reform to remove barriers to renewables and market distortions:

> *The reforms needed to address market barriers to renewables include:*
>
> - *Streamlined and uniform planning procedures and permitting systems and integrated least cost network planning;*
> - *Access to the grid at fair, transparent prices and removal of discriminatory access and transmission tariffs;*
> - *Fair and transparent pricing for power throughout a network, with recognition and remuneration for the benefits of embedded generation;*
> - *Unbundling of utilities into separate generation and distribution companies;*

- *The costs of grid infrastructure development and reinforcement must be carried by the grid management authority rather than individual renewable energy projects; and*
- *Disclosure of fuel mix and environmental impact to end-users to enable consumers to make an informed choice of power source.*

An important point is that fossil fuel and nuclear power sources still receive large subsidies, which distort the markets and increase the need to support renewables (see Box 12.4). Wind power would not need special provisions if markets were not distorted by the fact that it is still virtually free for electricity producers to pollute, according to the Greenpeace report.

Box 12.4 SUBSIDIES TO CONVENTIONAL POWER SOURCES

- The UNDP World Energy Assessment in 2000 stated that in the mid-1990s governments worldwide were subsidizing fossil fuel and nuclear power by around $250–300 billion per year.
- The World Bank estimated in 1997 that annual fossil fuel subsidies were $58 billion in the OECD and the 20 biggest countries outside the OECD.

Over the last three decades 92 per cent of all R&D funding ($267 billion) has been spent on non-renewables, largely fossil fuel and nuclear technologies, compared to 8 per cent ($23 billion) for all renewable technologies.

13
Wind Power and the Environment

Wind turbines use the renewable power from the wind; they don't produce any emissions or require any fuel transport that can harm the environment. A wind turbine pays back the energy that has been used to manufacture it in three to nine months, depending on the wind resources at the site, the size of turbine and the method of calculation (Elsam, 2004). The turbine can be dismantled without leaving any lasting traces behind, and most of the material can be recycled. All other ways to produce electric power in new power plants have greater impacts on the environment.

From an environmental point of view, wind power is the best option; it has a positive impact on the global and regional environment. The risks associated with climate change, acidification and eutrophication and their impacts on agriculture, forests, lakes, landscape and human health decrease with more electricity being generated by wind (see Table 13.1).

The concept of environmental impact encompasses a lot of different kinds of impact. Wind turbines can cause impacts on the environment by noise, shadow flicker, and changing views of landscapes, flora and fauna, and cultural heritage. But a positive impact is that the emissions from the power system are reduced. To be able to do reasonable assessments of these impacts, this concept has to be specified (Wizelius et al, 2005) (see Box 13.1).

Table 13.1 *Environment impact from different energy sources*

Energy source	Raw product	Emission	Other impacts
Combustion	Coal, oil, gas	CO_2, NO_x, SO_x, VOC, ash	Oil exploitation, mines, transport
Combustion	Biomass	NO_x, SO_x, VOC, ashes	Forestry, transport
Hydropower	Streaming water	None	Exploitation of land and watersheds
Wind power	Wind	None	Land use, noise
Solar heating, solar cells	Solar radiation	None	Land use

CO_2 = carbon dioxide, NO_x = nitrogen oxides, SO_x = sulphur oxides, VOC = volatile organic compounds

BOX 13.1 ENVIRONMENT IMPACT FROM WIND TURBINES

The impact on environment can be divided into the following categories:

Eco-system
Chemical/physical impact – acidification, eutrophication, climate change, pollutants, etc.

Health and comfort
Impacts on neighbours that can cause nuisance – noise, shadow flicker, safety.

Culture
Visual impact on landscape; cultural heritage.

Impacts on the environment can be *local*, *regional* or *global*. Burning of fossil fuels (coal, oil, natural gas) produces emissions of the greenhouse gas carbon dioxide, sulphur and nitrogen oxides, volatile organic compounds (hydrocarbons, etc.), heavy metals (lead, cadmium and mercury), as well as soot and particles. The extraction of fuel from mines and oil and gas wells also has serious local impacts on the environment, as well as causing further emissions. The transport of the fuel from source to the power plants requires energy and is yet another source of emissions.

The environmental gain that the development of wind power creates depends on the power system where the turbines are installed: how the power would be produced without wind power and how that would affect the environment.

Sweden, for example, belongs to a large north European power system. The power systems in the Nordic countries – Sweden, Norway, Denmark and Finland – are interconnected and there are also cables for power exchange with Germany and Poland. When a new wind turbine starts to produce power in Sweden, it replaces the same amount of power from a coal-fired plant in the north European power system. New wind turbines in Sweden do not replace power from hydrostations and nuclear reactors but imported power from a coal-fired plant (Holttinen, 2004). Thus by assessing the emissions from coal-fired plants, the environmental gains can be quantified (see Table 13.2).

The amount of reduction depends on what power plant is used for comparison in the calculation. CO_2 emissions are approximately the same from all coal-fired power plants; emissions of SO_x and NO_x, however, can vary. Table 13.2 is based on a coal-fired power plant with good equipment for emission reduction.

The amount of reduction not only depends on what types of power plants are installed in the power system where wind power is connected but also, of course, on the actual share of wind power in the power production – the wind power *penetration*. As the penetration increases, wind power will replace not only coal-fired

Table 13.2 *The contribution of wind power to reduction of environment impact in the Nordic power system*

Substance	1kWh	1GWh
SO_x	0.37g	370kg
CO_2	850g*	850 tonnes
NO_x	1.2g	1.2 tonnes

* 800–900g/kWh (Holttinen, 2004).

The yearly environmental gain of a wind power project can be calculated by multiplying yearly power production with the figures in the table. To this solid refuse in the form of dross (clinkers) of around 52g/kWh (52 tonnes/GWh) has to be added.

Sources: Danish Wind Turbine Owners' Association (2001) and Holttinen (2004)

power plants, but also some hydropower and nuclear power in the Nordic power system. With 4.3 per cent penetration (16TWh/a) CO_2 emissions are reduced by 700g/kWh (as a result of the wind power), with 12.2 per cent (46TWh) the reduction is by 650g/kWh (Holttinen, 2004).

In the Baltic power system, where a large share of the power comes from oil-shale-fired power plants, the reductions are larger, 1.05kg/kWh. However, to get an actual reduction of the CO_2 emissions, the penetration has to be quite large, since the fossil power plants only can be regulated in steps of 10MW. Small shares of wind power will only increase losses in the grid.

The reduction of CO_2 and other emissions depends on the design and characteristics of the power system where the wind turbines are installed; these figures have to be calculated separately for each power system. But in most power systems wind power will produce a significant reduction in CO_2 emissions.

Development of wind power contributes to reducing the negative impact power production has on the global environment, since carbon dioxide emissions that can change the global climate are reduced. It also contributes to reducing emissions of cross-border air pollution from sulphur and nitrogen oxides that cause acidification, eutrophication and other environment impacts at the regional level. In short, wind power has a *positive* impact on the global and regional environment.

Land demand

It might seem obvious that wind power needs more land than other energy sources. If you analyse this question and compare wind power with other kinds of power plants, however, it turns out to be less self-evident. Power plants that use fossil fuels or uranium use land areas in the whole chain of production, from the exploitation of the raw material to waste dumps – mines, oil wells, refineries, ports and storage facilities. Wind power's demand on land varies from 0.018 to 0.49ha/MW

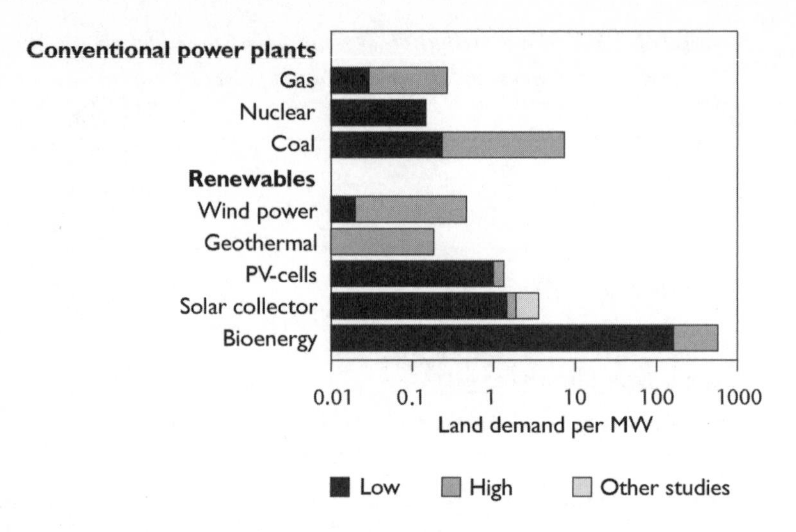

Figure 13.1 *Land requirement per MW*

Wind power does not necessarily need more land than other energy sources. Other power plants need fuel, which is taken from mines; the fuel has to be transported and for this harbours and other infrastructure are needed; and the remains – soot, ash, dross and radioactive waste – need large areas for storage. The diagram shows the land requirement in ha/MW for different types of power plants (note that the horizontal axis has a logarithmic scale). The land requirement for wind power varies from 0.018 to 0.49ha/MW for the five wind farms from the middle of the 1990s that are included in the study.

Source: Milborrow (1998)

including foundations, access roads, transformers and other equipment, according to an empirical study; a UK nuclear plant requires 0.16ha/MW according to the same study (see Figures 13.1 and 13.2).

According to the Swedish government report 'Rätt plats för Vindkraft' (SOU, 1999, p75) land requirement, defined as the area limited by a line surrounding the outer towers in a group of turbines, depends on the configuration. If the turbines are sited along a line, the land requirement is very small; if they are arranged in, say, three rows of four turbines, the land requirement increases. A group of twelve 1.5MW turbines in a 3×4 group require an area of 81ha, if they are sited in a 2×6 group they only need 47ha. The total nominal power of such group is 18MW, which gives land requirements of 4.5 and 2.6ha/MW respectively. However, 99 per cent of this area can still be used as before, so the actual land requirement is only 0.045 and 0.026ha/MW. And the better the wind resources are used, the fewer the turbines that have to be installed.

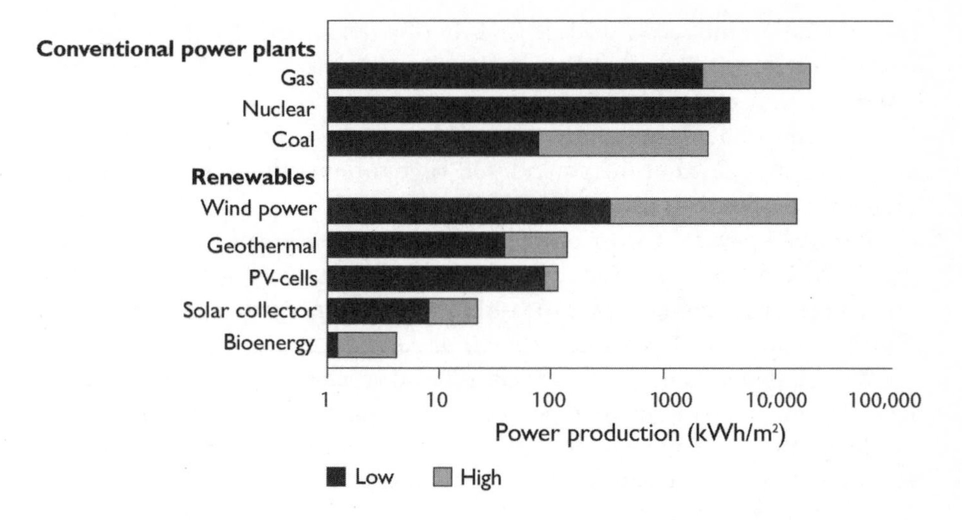

Figure 13.2 *Land requirement in relation to productivity*

If the productivity in kWh/m² is compared, wind power can still compete with fossil and nuclear power plants. With the large wind turbines installed in the 2000s the land requirement of wind power is actually less per kWh than other power plants as the areas between the turbines can continue to be used as before the wind power plants were installed.

Source: Milborrow (1998)

Local impacts

Wind turbines have to be sited in places with good wind conditions. Access roads sometimes have to be built to make it possible to transport the turbines to the site, and power lines have to be installed to connect the turbines to the grid.

The energy content of the wind increases with the cube of the wind speed, and a wind turbine sited at the coast can produce 25–30 per cent more power than a turbine 2km inland. The wind resource is a crucial factor for the economic viability of a wind power installation – ultimately it defines the areas where it will be feasible to exploit wind energy.

Wind turbines are usually mounted on concrete foundations, but can also be bolted into the ground if it consists of solid stone. A concrete foundation consists of a large, usually square, foundation cast in a mould a few metres below ground level and a concrete pole with a steel flange where the tower is bolted. The foundation is covered with earth and the foundation does not cover more area than the tower base – a few square metres at ground level.

An access road has to be built from the closest road so that excavator, mobile crane and other heavy vehicles can access the site if the ground doesn't have enough carrying capacity. These roads can be temporary, sometimes consisting simply of reinforcements of some sections or steel plates being laid on the ground.

The demand on the access road depends on the terrain and the size of the turbines. When the turbines are delivered, the access road has to be able to bear heavy lorries with trailers and a heavy mobile crane.

The transformer is built next to the turbine and is connected to the grid by a cable, usually buried in the ground. On large turbines the transformer is preinstalled in the tower or the nacelle.

The *direct physical impact* on the environment consists of the foundations, access road and cables to the grid, and the turbine itself, which will demand some of the air space at the site. Once the turbines have been installed, however, most of the land can be used as before, as arable or pasture land, for example.

Wind turbines do not cause any emissions that have impacts on the environment. The only impacts are that the rotor creates some noise, the turbines are visible during daytime and when the sun shines the rotor throws a rotating shadow that moves from west to east from sunrise to sunset.

Flora and fauna

The impact on flora and fauna depends on the types of vegetation and animal life in the area.

Vegetation can be affected during the building phase or by changes in the hydrological conditions due to the foundations, cable ditches, etc. This in normal conditions is seldom a problem. Concerning wildlife, the risk of impacts on birds has been debated and a lot of research has been done to clarify this issue.

In the 1980s there were problems in Denmark with small birds that built nests in the nacelles, so the manufacturers had to cover all openings with lattice. In the US many falcons collided with the rotor blades. The reason for this turned out to be that the smaller birds that were the falcons' prey built their nests in the towers. These turbines had lattice towers that had many perfect branch forks for bird nests. Covering the lattice towers with sheet metal solved this problem. In southern Spain several vultures were killed at the Tarifa wind farm. It turned out that the vultures gathered at a large waste dump in the middle of the wind farm and moving the dump to a better place solved the problem. The risk of birds colliding with wind turbines has been shown in practice to be relatively small.

In the Altamont pass in California, however, there are still problems with birds of prey colliding with turbines, and from August to December 2005 four dead sea eagles, two of them adults, were found close to the large 150MW wind farm, with 68 operational wind turbines, on the island of Smöla in Norway. The sea eagles had obviously been killed by the turbines (NINA, 2006).

It should be pointed out in this regard, however, that birds live a fairly precarious life in any case, with about 30 per cent dying during their first year due to collisions with natural or man-made objects (windows, high structures, power lines,

Table 13.3 *Bird mortality from various causes in the US*

Object	Mortality, million birds a year
Power grid	130–174
Cars and trucks	60–80
Buildings	100–1000
Telecom towers	40–50
Pesticides	67
Cats, domestic and feral*	39
Wind turbines	0.0064

* Wisconsin only.

Source: Sagrillo (2003)

etc.). In the US many scientific studies have been made to estimate the number of bird deaths a year from different causes (see Table 13.3).

Several studies have been conducted on bird mortality caused by wind turbines. The National Wind Coordinating Committee in the US analysed all such studies to the end of 2001 and estimated the total annual mortality from 3500 operating turbines in the US to be 6400 bird fatalities per year for all bird species. This would correspond to 0.01–0.02 per cent of the annual avian collision fatalities caused by man-made structures and activities in the US (NWCC, 2001). Thus the frequency of birds colliding with wind turbines has been shown to be relatively small.

Another impact on birdlife could be that birds are scared away from areas where wind turbines are erected. This scaring effect seems to vary between species, but most birds don't seem to be frightened by turbines and get accustomed to them quite quickly.

The impact offshore wind farms have on seabirds has also been investigated thoroughly. Comprehensive studies have been made at Tunö Knob and Nysted in Denmark and in the Kalmar Straight in Sweden, where a wind farm has been installed in the middle of the migration route, meaning that 1.5 million birds pass the 12 wind turbines each year. The routes of the migrating birds have been observed visually in daytime and by radar at night and during mist from 1999 (before the turbines were installed) to 2003. During this period only one collision was observed. The worst-case scenario of birds being killed by colliding with the wind turbines was estimated at 14 birds a year, which is considered negligible (hunters kill several hundreds of seabirds each year in the area, for example). The study shows that birds will spot and avoid the turbines in all kinds of weather.

Important resting habitats and nesting areas for birds are usually designated as bird protection areas. However, it is always important to investigate the bird situation in coastal areas and offshore and to adapt the wind farms to these conditions to minimize the impact.

Wind turbines have a technical and economic lifetime of at least 20 years, which can be prolonged by exchanging vital components such as gearboxes, generators and blades. The foundations have a much longer lifetime and could potentially be reused for a new installation at the same place, although the fast technical development of turbines makes this less likely. Wind turbines can be dismantled in a day and the site can be restored to its original state. Most of the components can be recycled. In short, wind turbines do not make a lasting impact on the environment.

Sound propagation

Wind turbines can cause two different types of noise: mechanical noise from the nacelle (gearbox, generator and other moving mechanical components) and aerodynamic noise from the rotor blades. There are carefully specified rules and methods for how noise from turbines is to be measured, how manufacturers are to specify the noise produced by their turbines, how the noise immission is to be calculated at different distances from a turbine, and what sound levels are to be permitted around different types of buildings.

The main distinction concerning measurements of noise is made between sound *emission* and sound *immission*. The sound emission is the sound that the turbine emits. The value for sound emission that the manufacturers declare, and that is used to calculate sound levels at different distances from the turbine, is the sound emission from the centre of the rotor when the wind speed is 8m/s at 10m above ground, for a turbine sited in an open landscape with roughness class 1.5 (roughness length 0.5).

The sound immission is the value that is measured (or calculated) at a specific distance from the turbine. If the sound emission and the hub height are known, the sound immission at different distances from the turbine can be calculated (see Figure 13.3).

The sound is measured in dBA (decibel A), which is an A-weighted sum of sound with different frequencies, adapted to the human ear's sensitivity for sound of such frequencies. The unit dBA gives a measure of the sound that the ear registers. Normal speech has a sound level of 65dBA, a modern refrigerator 35–40dBA, a city street about 75dBA, a discotheque around 100dBA and a quiet bedroom 30dBA.

The sound emission from a wind turbine is about 100dBA, varying from 95 to 108dBA. The sound from modern wind turbines all comes from the rotor blades: the mechanical noise has been eliminated by sound-absorbing materials in the nacelle, better precision in the manufactured components and damping. Nowadays mechanical sound can be heard only when a component begins to fail. The sound from the rotor is an aerodynamic swishing sound. Neither infra- nor ultra-sound, in other words sound with too low or too high a frequency to be

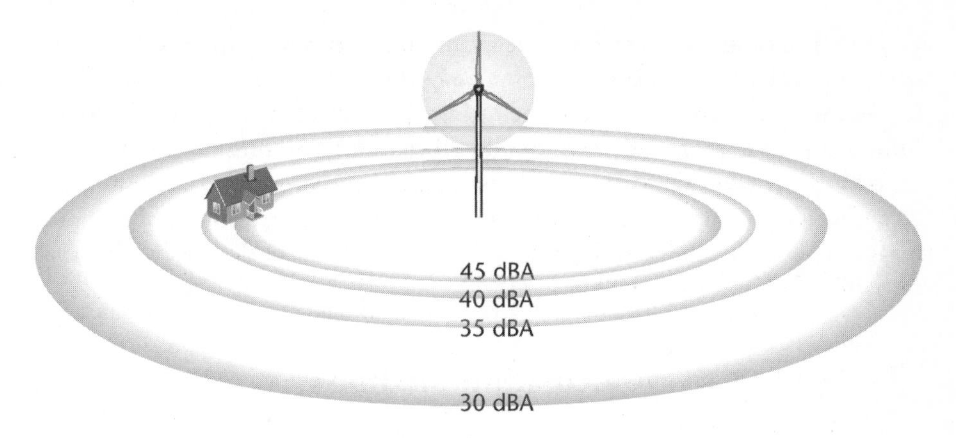

Figure 13.3 *Sound propagation from a wind turbine*

Sound immission at different distances from a wind turbine.

Source: Typoform

registered by the human ear, has been registered from wind turbines (Remmers, 1998).

There are no significant sound differences between small and large turbines, since the sound level depends on the tip speed, which is about the same for most turbines (small turbines have a higher rotational speed than large ones). Manufacturers have, however, managed to decrease this swishing sound during years of continuous development, by changing the form of the blades. The level of sound emission from a turbine is decisive for the permissible distance to neighbouring houses (see Table 13.4).

The sound from wind turbines differs from other kinds of industrial noise. In modern turbines noise from the machinery has been eliminated. The aerodynamic swishing noise has the same character as the rustling of leaves or other wind-induced noise. Turbines with variable speed rotate slower in low winds and the noise level is lower than the background sound level at almost all wind speeds.

Table 13.4 *Sound level from wind turbines/distances (m)*

Immission / Emission	45dBA	40dBA	35dBA
105dBA	350m	575m	775m
100dBA	200m	350m	575m
95dBA	120m	200m	350m

The sound emission from a wind turbine (given in the technical specifications for the turbine) is usually in the range 95 to 105dBA. The table shows rounded values to give an idea of appropriate distances and how they differ for various emission values. The decibel scale is logarithmic: an increase by 3dBA corresponds to a doubling of the sound pressure (power).

Wind turbines can in fact only be heard under certain conditions. When the wind wanes the turbines stop and can't be heard at all. When the wind speed is higher than 8m/s the sound from the turbine is drowned by background sounds from rustling leaves and other wind-induced sound. Wind turbines will be heard only when the wind speed is between cut-in wind speed, 3–4m/s, and 8m/s; the recommended values for maximum sound level are reached only at 8m/s at a height of 10m. The sound will spread more on the lee side of the turbine. In other directions the sound level will be lower.

Calculation methods

The sound immission at different distances from a wind turbine can be calculated with models for sound propagation. There is an international standard method for this, ISO9613-2, but unfortunately the authorities in some countries have decided to use their own models instead. Their acoustic experts claim that their models are more accurate for wind turbines than the international standard model. However, the results from these different models do not differ very much (see Table 13.5).

The results 500 metres from a wind turbine differ by less than 2dBA. Considering that the smallest difference in sound level that the human ear can perceive is 3dBA, the results are in practice the same. For the minimum distance to a dwelling, where the maximum sound immission is 35dBA, the distance will be a little more than 500 metres in the Netherlands, Denmark and Sweden and slightly less than 500 metres under the international standard used in Germany and other countries. Close to the turbine the international standard is 'stricter', while the difference at 40dBA is only 20 metres (Wizelius et al, 2005) (see Table 13.6).

Table 13.5 *Sound immission according to different calculation models**

ISO9613-2	Denmark	The Netherlands	Sweden
34.1	35.5	36.0	35.5

* for a wind turbine with 50m hub height and a sound emission of 100dBA; sound immission 500m from the turbine (dBA).

Table 13.6 *Distance to 35, 40 and 45dBA*

dBA	ISO9613-2	Denmark	The Netherlands	Sweden
35	465	525	555	525
40	305	325	325	325
45	205	195	185	195

Rules and regulations

Rules for which sound immission are recommended for neighbouring houses differ in different countries. In Denmark the limit for dwellings is 45dBA, in Sweden the limit is 40dBA. In the UK the sound immission may not increase by more than 5dBA above the background noise level (see Table 13.7).

Table 13.7 *Recommended limits for sound immission in different countries (dBA)*

Country	Work areas – office, industry	Dwellings	Villages, farms	Recreation areas
Denmark	–	45	45	40
Germany	50–70	40	45	35
The Netherlands	40	35	30	–
UK	+ 5*	+ 5*	+ 5*	+ 5*
France	+ 3*	+ 3*	+ 3*	+ 3*
Norway	40	40	40	40
Sweden	50	40	40	35

* max increase from background noise during evening and night.

Shadows and reflections

During some periods of the day wind turbines can create shadows and reflections that can be disturbing if the turbines are unsuitably sited in relation to neighbouring buildings. The problem with reflections from rotor blades has been eliminated already, since the blades on modern turbines have an anti-reflection coating. The rotating shadows from rotors can also create a stroboscopic effect when they pass a window, which might be an unpleasant surprise if the risk hadn't been considered before the turbines were installed.

The risk of being disturbed by shadows is greater the closer a house is to a wind turbine. However, due to rules of maximum sound immission, the minimum distance to the closest neighbour is usually six to ten rotor diameters, and at that distance shadows will occur only during some short periods each day in limited periods of the year. A shadow will also be 'diluted' with distance: its sharpness decreases and it finally disappears due to optical phenomena in the atmosphere.

Theoretically a shadow from a turbine can reach 4.8km (for a turbine with a 45m rotor diameter), which would occur just after sunrise and just before sunset. In practice, though, a shadow will have a maximum reach of 1.4km (for a 2MW turbine with 2m blade width) although shadow effects are calculated for a distance of 2km.

The shadow from a wind turbine moves in the same manner as the shadow of a sundial, from west through north to east from sunrise to sunset. Since the sun

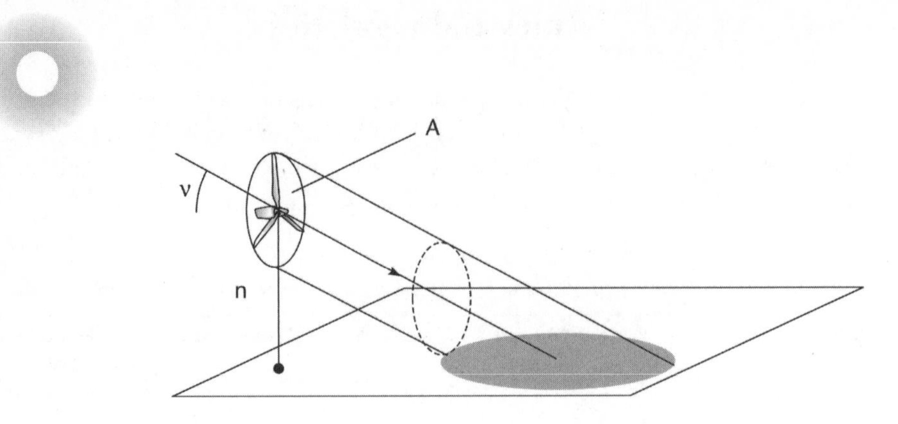

Figure 13.4 *Model for shadow calculation*

Using the hub height (*n*), the rotor area (*A*) and the inclination of the sun in relation to the horizontal plane (*v*), the location of the shadow can be calculated.

Source: Typoform/EMD (2005)

will rise later in the winter, when the sun's altitude also will be lower, the shadow will move along different paths in different seasons. Since the altitude of the sun can be calculated exactly for each time of the day and for different latitudes, the path of the shadow at each place can be exactly calculated (see Figure 13.4).

A house situated due west of a wind turbine can get shadow flicker at 6 o'clock in the morning, a house north of a turbine at noon and a house east of a turbine at 6 o'clock in the evening. The shadow is much shorter in the summer than in the winter, and if houses are 500 metres away from the turbine none will get shadow flicker for more than two short periods of the year, and this for a maximum of 20 minutes per day. During what time of the day a shadow from a wind turbine will fall in a specific place and for how many hours a year this may occur can be calculated using a web-based calculator that is available on the website of the Danish Wind Industry Association: www.windpower.org.

The results of these estimates or calculations show the theoretical maximum time a house can get shadow flicker from a turbine. It shows the 'worst' case, which is if the sun is always shining and the wind always blowing from a direction that gives maximum shadow impact (rotor perpendicular to the window). Since the sky is sometimes overcast and wind speed and direction varies, the real time for shadow flicker is much lower, less than one third of the worst case. If information about sunshine and the distribution of wind directions per month are available, the time for actual shadow impact can be calculated.

The calculation model in Figure 13.4 is based on simplifications which give an overestimation of the shadow impact. In this so-called geometrical model the sun is reduced to a dot and the light/shadow spreads in a vacuum. In reality, however, the sun covers three degrees of the sky, so 'sunbeams' actually meet behind the rotor blades and wash out the shadows at a certain distance. Furthermore, in reality the air also diffuses the light. These two phenomena have been studied by the German scientist Hans-Dieter Freund, who has introduced a physical correction factor to get results that better reflect reality.

The maximum distance for a shadow to be visible depends on the hub height and rotor diameter of the wind turbine. The length of the shadow also varies with the opaqueness of the air, which depends on humidity and temperature. On a clear day in winter the shadow can be much longer than on a clear summer day. The shadow will also be visible at longer distances on a vertical than on a horizontal surface. According to Freund the maximum length of shadows can be calculated as shown in Table 13.8.

Table 13.8 *Maximum length of shadows from wind turbines*

		Summer		Winter	
Hub height (m)	*Rotor diameter (m)*	*Horizontal*	*Vertical*	*Horizontal*	*Vertical*
25	25	200m	350m	300m	700m
50	50	300m	700m	600m	1250m
75	75	500m	1100m	850m	1800m
100	100	600m	1375m	1100m	2300m
125	120	700m	1650m	1300m	2700m

The maximum distance for a shadow on a vertical surface, like a window or façade, for wind turbines with 75 metre hub height and the same rotor diameter, is 1100 metres in the summer and some 700 metres more during the rare winter days when the air is crystal clear. The size of the shadow, the area it sweeps, will also decrease with distance.

Source: Freund (2002)

Visual impact on the landscape

Wind turbines are visible objects and consequently have a visual impact on the landscape, like most other structures, factories, power lines, roads and so on. Since wind turbines are tall and have a rotating rotor, they attract attention of passers by. Therefore wind turbines are considered to have a comparatively large impact on the impression of a landscape. Whether this impact should be considered positive or negative is a matter of individual opinion and is also influenced by the type of surroundings. And opinions do differ: some consider wind turbines ugly machines that turn the environment into an industrial area; others view them as slender sculptures that visualize the power of the wind, or as a clever and therefore acceptable way to use the power that nature offers for free.

Farmers usually want to use available natural resources in an efficient way and consider the surroundings as a production landscape. They also have to make a living out of them. Tourists and summer cottage owners often view the land-

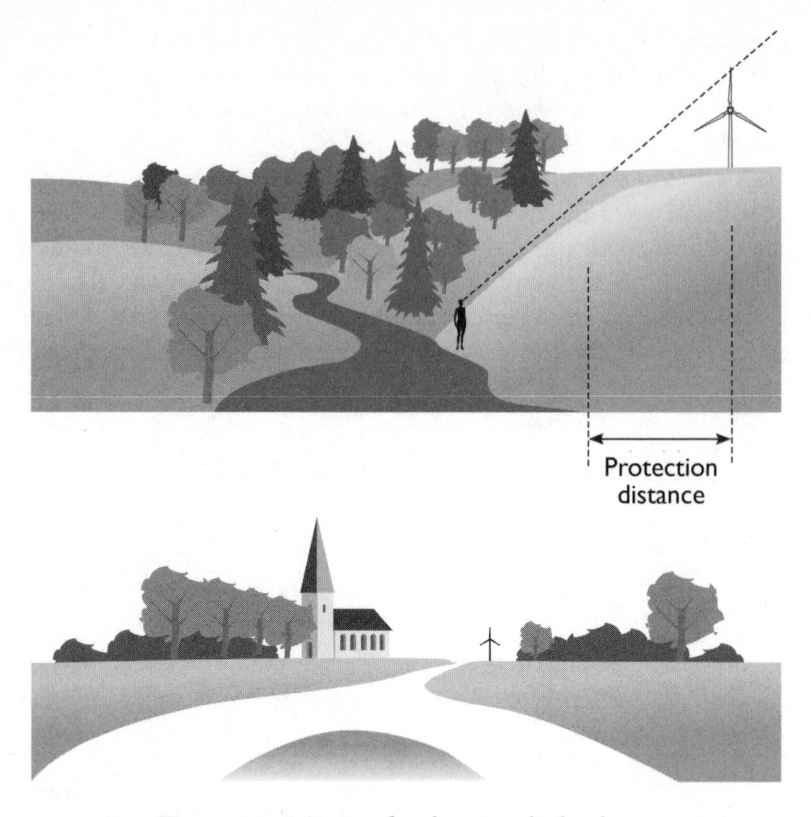

Figure 13.5 *Siting of turbines in the landscape*

It is important to consider from what places a planned wind turbine installation will be visible and what impact the turbines will have on the experience of existing buildings and nature. This can be done by checking the views from different viewpoints. In a valley the hill brows should be kept free, and distances to churches and other similar buildings should be long enough so that the turbines don't destroy the impression of them.

Source: Typoform/Länsstyrelserna i Skåne (1996)

scape as a 'postcard' – there to give aesthetic and similar experiences. Furthermore, opinions and experiences tend to change over time, so after some time turbines can come to be considered natural and valuable components in the landscape.

It is difficult to determine how wind turbines affect the landscape, since the experience of the values of a landscape is subjective and people can have quite different opinions about them. There certainly are some turbines that are sited in bad places, as well as turbines that suit the landscape well. Obviously there are some factors that make a difference. However, it is very hard to formulate general rules for the best way turbines should be sited in a landscape (see Figure 13.5).

A developer who plans to install wind turbines has to choose the site very carefully to make the turbines produce as much power as possible. The area that

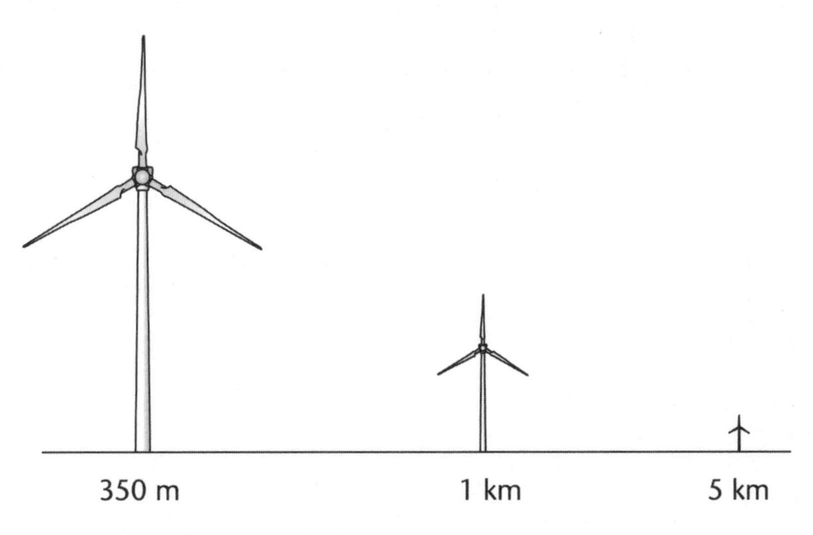

Figure 13.6 *Visual impact of wind turbines*

Within a distance of 350m (10 hub heights) a wind turbine (with a 35m tower) dominates the landscape, 1km away the turbine is clearly visible but not dominating, and 5km away the turbine will be perceived as a part of the landscape.

Source: Typoform/Gipe (1995)

surrounds the turbines should be as open as possible, with a proper distance to buildings, groves and other obstacles that can disturb the wind. Usually this is also a good site from an aesthetic point of view.

The visual impact of wind turbines decreases quickly with distance. A limit for when the visual impact can be considered negligible can be set: a common rule of thumb is that wind turbines dominate the landscape within a distance of ten times the turbines hub height, in other words within a circle of 600m radius for a turbine with a 60m tower. This will be outside the 40dBA limit for sound immission; further away the turbine is clearly visible up to 5km and after that it melts into the landscape (see Figure 13.6).

A wind turbine will be visible up to a distance of 400 times the hub height, in other words up to 20km away for a turbine with 50m hub height. However, turbines will melt into the landscape after about 5km, the distance when this occurs depending on the turbine size and the character of the landscape.
The Swedish government report 'Rätt plats för Vindkraft' (SOU, 1999, p75) makes a different classification, however (see Box 13.2).

The impact on the landscape of a planned wind power project can be illustrated with photomontages. Most wind power software programs offer this option, but it can also be done with Photoshop or similar software.

The number and size of the turbines and their configuration also influence the visual impact. Experience has shown, though, that it is very hard for the human eye to distinguish between small and large turbines. The eye will interpret

BOX 13.2 VISUAL ZONES FOR WIND TURBINES

Close zone
Up to 2–3km: wind turbines are a dominating element.

Intermediate zone
From 3 to about 7km: the visibility of the turbines varies according to the character of the landscape. In open landscapes with a wide view and unbroken ground (flat terrain) the turbines are clearly visible, although it can be difficult to perceive their size. If the landscape is cultivated with many groves, forested areas, building, etc., the visibility is reduced.

Distant zone
Up to about 12km: even at these distances wind turbines will be clearly visible in an open landscape with an unobstructed view. However, the forms of the landscape and the elements in it will reduce the dominance of the turbines.

Very distant zone
More than 10–12km: in a landscape with an unobstructed view wind turbines can be perceived as small objects on the horizon, but it can be difficult to distinguish them from other objects in the landscape.

a difference in height to a difference in distance. It is also hard to note different patterns in a wind farm.

It is not possible to assess how the values of a landscape will be affected by new structures or other changes just by reading maps or by interpretations of laws and regulations. These values are defined by the evaluations made by the people who work and live in the landscape. The value of different sites and views depend on the traditions, memories and feelings of the local inhabitants. These aspects are usually clarified during the public consultations that are a part of the planning process.

Finally it is important to realize that the landscapes we live in are created by man. There is no natural landscape. This is the view of modern geographers:

> *Landscape is not scenery… it is really no more than a collection, a system of man-made spaces on the surface of the earth. The natural environment is always artificial.* J. B. Jackson, quoted in Pasqualetti et al (2002)

A landscape in a sustainable society will look different from the landscape of today.

14
Wind Power Planning

Wind power planning can mean several things and be undertaken at different levels by various actors. To start from the top we have international treaties, or recommendations like the Kyoto Protocol. To achieve the targets for reductions of CO_2 emissions, to address the threat of global climate change, an important step is to replace fossil fuels with renewable energy sources, one of which is wind power. These kinds of international treaties govern overall ambitions for development of wind power and other renewable energy sources at the national and international levels.

Most countries also have some kind of national energy policy, of which development of wind power can be a part. These could set a target for how much and how fast wind power should be developed at the national and sometimes also the regional level. An important part of this policy is the set of laws, rules and regulations that are applied to wind power development. These aspects were discussed in the preceding two chapters.

Spatial planning at the local and regional levels is governed by local and regional authorities, which handle building permissions, environmental permits and so on. In some counties and/or municipalities comprehensive plans have been worked out whereby suitable areas for wind power development are designated.

On the other side of coin there are the developers, who make their own plans for wind power development. Companies have general business ideas, business plans, strategies and ambitions that result in area surveys for suitable sites for wind power, feasibility studies and project development programmes. This kind of planning is described in Part V, Project Development.

In this chapter the focus is on spatial planning. A policy framework – rules and regulations that stimulate or act as barriers to wind power development – regulates this planning. The legislation for planning differs in different countries; since I am most familiar with the situation in Sweden and Denmark, most of my examples will be drawn from these countries. Since wind power development has been very successful in Denmark but very slow in Sweden, a comparison will also produce some ideas on how to implement an efficient planning strategy and to avoid planning methods that create barriers.

Targets for wind power development

In Denmark the government has set targets for wind power development, the first in the government energy strategy Energy 81, which set a target of 1000MW of wind energy by 2000. This was followed by the action plans Energy 2000 in 1990 and Energy 21 in 1996, which set a target of 1500MW by 2005, a target that was already reached in 1999. A long-term target for 2030 was also set at 5500MW, of which 4000MW were expected to be installed offshore. In achieving this target, 50 per cent of the electric power in Denmark will be supplied by wind power.

In 1992 a Wind Turbines Act entered into force. The government required municipal and regional authorities to make plans for wind turbine siting through a planning directive. Although no quotas were set, most counties managed to select sites with good wind resources through extensive public consultation with local residents. More than 2600MW of capacity were identified.

Spatial planning in Denmark is performed at three levels: national planning by central authorities, regional planning by the counties and local planning by the municipalities. The Spatial Planning Act requires counties to designate areas in their regional plans for new energy projects. In response to the regional plan municipalities develop local wind turbine plans, which prescribe areas where wind turbines can be installed, for single turbines, clusters or wind farms, as well as conditions such as hub height, colour and so forth.

Counties then issue zoning permits and installation permits to the municipal plan according to the act. Each county sets up guidelines for regional planning which prescribe the terms for wind turbine installations in the county. Spatial planning also emphasizes the importance of involving grid operators so that they can prepare for the expansion by strengthening the grid and the new turbines can be smoothly connected to the power system (see Box 14.1).

The Environment Protection Act also regulates siting of wind turbines in Denmark. This act prescribes proximity guidelines such as distances from turbines to the coastline (300m), lakes and streams (150m), forests (300m), ancient monuments (100m) and churches (300m). No other specific permissions were needed to install wind turbines on land. However, for larger installations an EIA has to be made as well.

Swedish planning target

In Sweden the government proposed a target for electric power production from renewable energy sources (wind power, bio-fuelled CHP and small-scale hydropower) in 2002. The target was set at 10TWh/a in 2010, with a target for wind power to reach 10TWh/a by 2015 (Swedish Government, 2002). These proposals were adopted by parliament in June 2002.

BOX 14.1 MUNICIPAL PLANNING IN THISTED MUNICIPALITY IN DENMARK

Thisted municipality on the Limfjord in Jutland has very good wind resources. When work with the municipal plan started, there were already 99 turbines online, most of them < 100kW. A work group was formed, with representatives from local interest groups and organizations, such as the nature protection association, the farmers' association, hunters, the local folklore society, the wind power association and utilities, as well as civil servants from the municipality and the county.

The starting point was the map of the municipality and the wind resource map. First, zones exempted in the regional wind power plan of the county were excluded: a large zone along the coast, minimum distances to buildings, etc. Then all areas that were considered possible for wind turbines were drawn into the map. All these sites were examined and areas unsuitable due to local obstacles, etc., were deleted.

The work group then sat around a table with the map and discussed and negotiated until a consensus was reached on all areas that would be included in the plan. This plan was then forwarded to the building committee, which examined the proposed areas and made some slight changes. Then the plan was exhibited to the public before the politicians in the municipality adopted it.

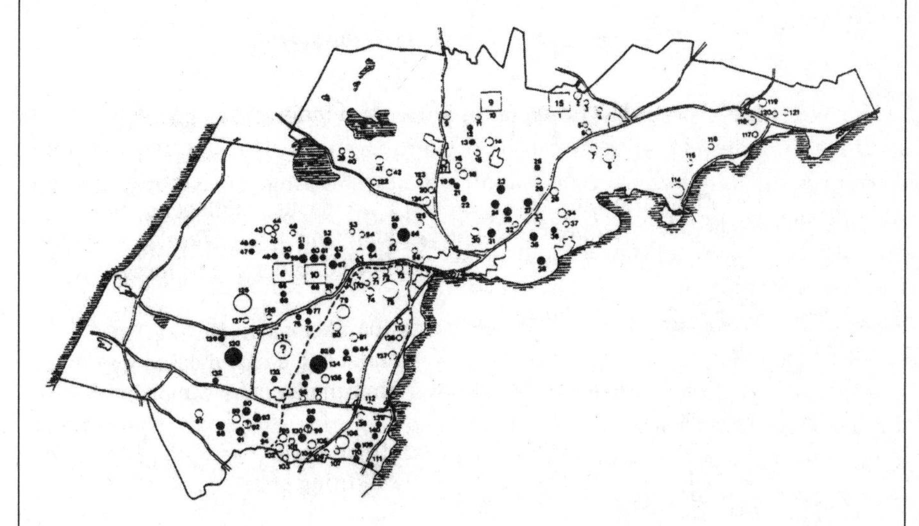

Figure 14.1 *Wind power plan in Thisted municipality*

When farmers, the nature protection society, the utilities and other interests had gained a hearing, the plan could still accommodate 108 wind turbines on the sites with filled spots. The empty spots had been deleted.

Source: Wizelius (1993)

There is, however, no specific target for wind power development in Sweden or how large a share of the electric power should be produced by wind turbines. The planning target just means that the municipalities should designate areas for wind power development in municipal comprehensive plans or similar documents to make it possible to produce 10TWh/a by 2015. Whether this will be realized depends on the economic conditions and other rules that politicians apply to wind power development.

In Sweden the municipalities have a *planning monopoly*. This means that it is the municipalities (not the counties or the central government) who have the exclusive right to decide how land and water areas within their borders shall be used. Guidelines for land use have to be published in *municipal comprehensive plans* (MCPs), which all municipalities are obliged to make. The municipalities have to consider different kinds of public interest, however (the county administration has to check that this is undertaken). One kind of public interest concerns so-called areas of national interest, specified in the Swedish Environment Act. The cultural heritage act, with its rules and regulations to protect and maintain buildings and areas of special interest, can also have an impact on the planning and permission process for wind power.

Areas of national interest

In the Swedish Environment Act areas of national interest have been defined and specified on maps. These areas can be of national interest for nature protection, recreation, the military, etc., or at sea for fisheries, shipping, etc. According to this act land and water areas that are especially suitable for certain types of exploitation, for example wind power plants, shall be protected from measures (in other words exploitation) that can come in conflict with the utilization of such plants. Areas of national interest should be protected so that they can be used for the specified purpose.

When these laws and regulations about areas of national interest were introduced, wind power had not entered the stage, so no areas for wind power were defined at that time. However, in 2004 the Energy Agency, in cooperation with the county administrations, added areas of *national interest for wind power* to the map.

The planning target of 10TWh/a by 2015 has also been portioned out to counties according to their wind resources and share of power consumption. However, no efforts have been made to take the next step, to specify targets for the municipalities. The areas of national interest for wind power have been specified at the regional level by the counties, but to be of any practical use these also have to be incorporated in the MCPs of the municipalities. The municipalities, however, have not been involved in this planning process, and no directive has been given to them to implement the targets at the local level.

At the national level central authorities have made a feasibility study using a geographical information system (GIS), which has resulted in a map. On this map areas where wind turbines should *not* be built ('stop-areas') and where there are strong opposing interests have been designated. This can be defined as *negative planning*.

The criterion to identify so-called stop-areas has been that several areas of national interest overlap (for example, nature protection, recreation and the military). With this method large areas of Sweden have been excluded from wind power development (see Figure 14.2).

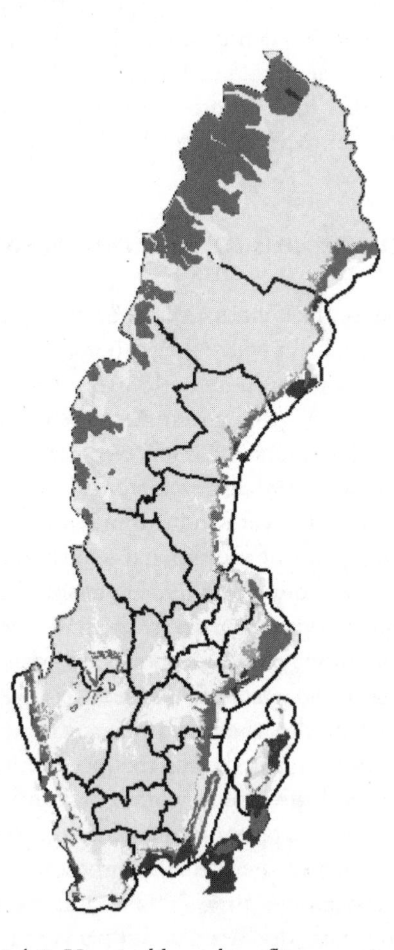

Figure 14.2 *Unsuitable and conflict areas in Sweden*

The dark grey areas on the map are defined as stop-areas and the light grey as areas with strong conflicts. This kind of negative planning – with general classification of areas – is made on too abstract a level. There are several large and also uncontroversial wind farms online within these stop-areas.

Source: Boverket (2002)

When this map is compared to reality, it turns out that there actually are several wind farms installed in stop-areas, for example on Gotland and in Kalmarsund (offshore). These projects have been quite uncontroversial and have been processed with comprehensive consultation with both authorities and the public. This indicates that a planning method using general criteria to exclude areas for wind power development isn't feasible: it acts as a barrier instead of a driver for wind power development.

To serve as a driver for wind power development, planning at the national level, for territories on land and offshore, has to be *positive*; in other words areas where, from a public and national policy point of view, it will be suitable to develop wind power should be identified and designated. On land this kind of planning is hard to conduct at a national level, since there are many local circumstances to consider, whereas for offshore planning it is entirely feasible and has been done in Denmark, Germany and the UK (see Figure 14.3).

Regional and municipal planning

Many wind power projects have been developed in regions and municipalities where there are no plans to direct where they should be sited. In Denmark several hundred wind turbines were installed in the 1980s without any plans. Spatial plans were made first in 1992–1994 at the government's request. In Sweden, Germany and most other countries development has followed the same pattern. The demand for planning seems to appear first when development has reached a level where politicians consider it necessary to direct wind turbines to the most suitable sites.

From the developers' point of view spatial wind power plans have both advantages and drawbacks. Without a plan that expresses a political intention about the suitability of different geographical areas, the outcome of each application is uncertain. The risk of spending time and money on projects that will be denied permission, and thus never generate any income, is higher when there are no plans. On the other hand it takes time, two years or more, for a municipality or region to actually work out a plan. During this period, from when the decision to make a plan is made until it has been worked out and politically approved, it is usually impossible to get any applications processed at all, since these decisions will depend on the outcome of the planning procedure. In Denmark development slowed down considerably during 1992–1994, but since then it has been running faster. In the long run local wind power plans should be an advantage for developers too.

Municipalities that decide not to make a wind power plan give up the opportunity to direct wind turbines to areas that, from a political point of view, are most suitable for this purpose. To make a usable wind power plan, the civil servants who undertake this need some special competence and a basic understanding of what criteria to use when choosing areas for this purpose. Otherwise there is

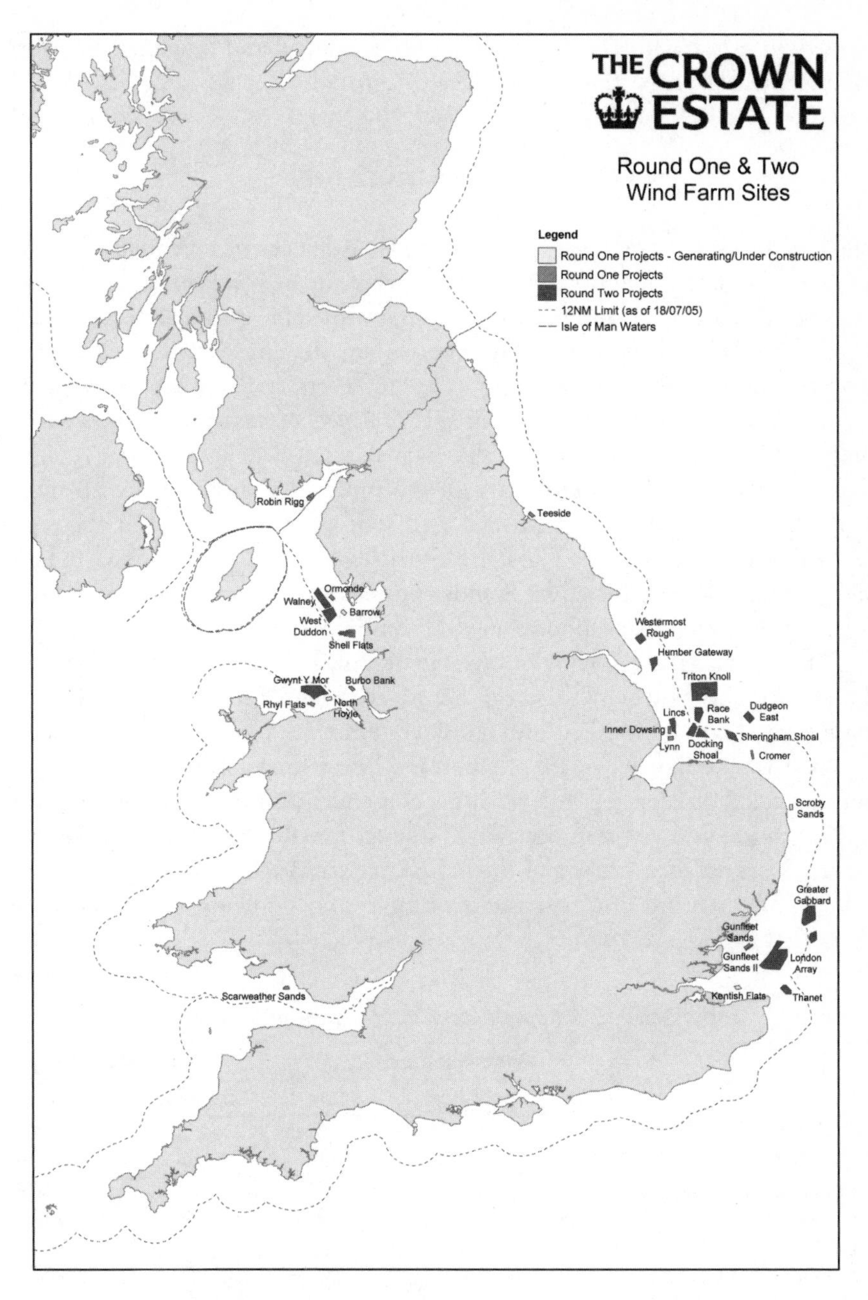

Figure 14.3 *Plan for offshore wind power in the UK*

On this map areas for wind power development offshore have been designated. The smaller areas close to shore are for round one, and will be developed first, the larger areas further from shore will be developed when round one is completed. Developers make tenders for development within these areas.

Source: Crown Estate (2003)

an obvious risk that areas where wind resources are bad and roughness very high, areas that will never be exploited, will be designated.

Planning methods

There are many very good planning tools available that can be used to develop wind power plans. The most important input for such plans are wind resource maps. Such maps can be developed with proper software if wind data are available for the area. In many regions and countries the authorities have developed such maps. Relevant geographical data for use in GIS software are also necessary. However, when preparing such maps for wind power planning it is always necessary, and important, to compare GIS data and maps with reality by making surveys in the field.

In Sweden three pilot projects with the aim of developing a good planning method for wind power were carried out. One of these resulted in the report 'Wind Power Planning in a Coastal Municipality: Case Study Tanum' (STEM, 2001), which was conducted by Scandiaconsult in Gothenburg and consisted of 100 A3 pages, with maps, photos and photomontages in colour.

The report starts out with an investigation of the available wind resources in the municipality and how these can be utilized when relevant laws and opposing interests have been taken into consideration. The whole archipelago area at the coast was excluded, as were other areas where strong conflicts with opposing interests could be expected. After that the consequences for nature, cultural heritage, recreation and visual impact were assessed. Finally these consequences were balanced against each other and three different scenarios for wind power development were worked out, with some offshore developments also included (see Table 14.1 and Figure 14.4).

Table 14.1 *Wind power scenarios for Tanum municipality*

	Installed capacity (MW)		Production (GWh/a)	
	additional	**sum**	additional	**sum**
Present (year 2000)		**7**		**11**
Zero-option	32	**39**	54	**65**
Development 1st stage	248	**287**	464	**529**
Development 2nd stage (including 1st stage)	255	**542**	664	**1193**
Development 3rd stage (incl. 1st and 2nd stage)	450	**992**	1183	**2376**
Option 1st stage	495	**534**	1253	**1319**
Electric power demand in Tanum (1991)				**115**
Electric power demand in Västra Götaland (county)				**ca 20,000**
Electric power demand in Sweden (1999)				**143,900**

In the zero-option wind turbines that at the time had received permission but had not yet been installed are included.

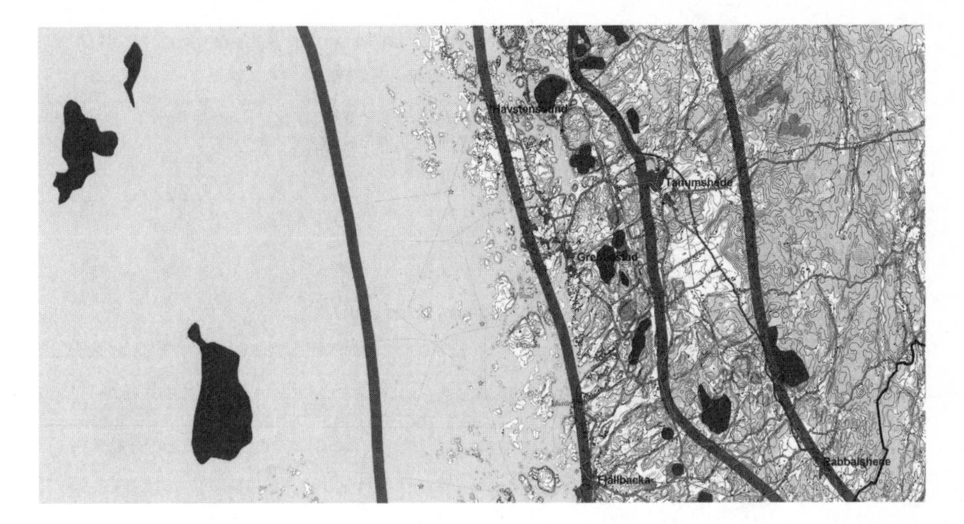

Figure 14.4 *Wind power areas in Tanum municipality*

The map shows areas according to development stage 1 plus offshore developments according to stages 2 and 3. This scenario gives three times as much power production than stage 1, about 1.4 TWh/a.

Source: STEM (2001)

Tanum municipality's share of the Swedish planning target for wind power (10TWh/a) is 221GWh/a, if we use criteria based on the region's power consumption and how good the wind resources are. Tanum has 0.22 per cent of the area but 2.1 per cent of the wind resources (more than 2800kWh/m^2/year at 50m agl) available in Sweden within its borders.

The report shows that development of just the first stage will give twice as much wind-generated power than the municipality's share of the national planning target. If all stages are developed, wind power could produce ten times as much. This implies that Sweden actually could raise the planning target to 100TWh/a, which is far above any realistic visions for the future. The highest target that has been mentioned is 30TWh/a, about 20 per cent of total power production. An important conclusion from this report is that availability of land (and water areas) is no limiting factor for wind power development in Sweden. Compared to countries like Denmark and Germany, wind power 'density' is very low (see Table 14.2).

The Tanum report demonstrates a useful and objective method where a GIS has been used as a planning tool. It is a good example of how well-founded background material for the decision-making process for a comprehensive municipal plan for wind power can be developed. It gives a good overview of the actual wind power potential that can be utilized. However, a comprehensive municipal plan is a political document that has to be processed by the politicians of a municipality.

Table 14.2 *Wind power in relation to area and population in different countries, 2004*

Country	Power [MW]*	Population [million]	kW/capita*	Area [km²]	kW/km²
Germany	14,609	82.5	0.18	356,733	41
Spain	6202	40.7	0.15	504,782	12
Denmark	3115	5.4	0.58	43,094	72
The Netherlands	912	16.2	0.06	41,863	22
Italy	891	57.3	0.02	301,300	3
Sweden	442	9.0	0.05	449,964	1

* The amount of wind power is given as installed power. Since there are wind turbines of different sizes there is no simple formula to convert installed power to number of turbines. The number of turbines is usually larger than the installed power; in Sweden there are about 700 turbines, in Denmark around 5000.

This study has so far not been transformed into a municipal plan, and the municipality, as well as the county, has a much more restrictive view on wind power development.

Wind power planning on Gotland

The island of Gotland in the Baltic Sea is the municipality in Sweden that has the most installed wind power. In 2005 it had 160 wind turbines connected to the power grid. The first modern wind turbine was installed in 1983 at Näsudden on the southwestern coast of Gotland. This was the prototype Näsudden I, with 2MW nominal power, a 70 metre concrete tower and a two-bladed rotor. It was installed on the wind power development area of the Vattenfall state utility.

In 1990 the first wind power cooperative in Sweden installed three 150kW turbines in the same area. In the coming years many new cooperatives were started and there are now ten cooperatives with 1458 members/part-owners that have invested SEK40 million (about €4.3 million) in wind power. When the market for cooperatives was saturated in the middle of the 1990s, farmers and companies started to invest in their own turbines. Most of the turbines were installed at Näsudden, where there now are 81 turbines online (see Figure 14.5). Wind turbines also spread to other parts of the island – single turbines and small groups in farming areas and two quite large wind farms at limestone quarries on northern Gotland.

Although Gotland has been in the front line of wind power development in Sweden, it was not until the end of the 1990s that the municipality worked out its first spatial wind power plan. Up to that time applications had been processed on an ad hoc basis, with permissions granted if the projects did not violate any

Figure 14.5 *Wind turbines on Näsudden, Gotland*

The wind turbines on Näsudden, Gotland, were installed during the 1990s. Each year a number of single turbines or small clusters have been added. Today there are turbines of many different sizes and brands. No overall planning has guided the development.

Source: Tore Wizelius

laws and regulations. The municipality had a very positive attitude towards wind power, and most projects were smaller than 1MW, so permissions from the county administration were not necessary.

The wind power plan was made in the form of a comprehensive municipal plan, with areas for large wind power projects on the southern part of Gotland added to the existing plan. The municipality formally adopted this plan in 1999 (see Figure 14.6).

In the plan it is stated that:

> *The principle is that development of more than 10 new wind turbines shall be done within the five new areas (areas 1–5) or at locations for large wind power installations on the plan map.*
>
> *The strategy is that the plan map only shows large coherent development areas to which the main part of future development should be directed. For developments outside these designated areas, it is up to the developer to find sites for wind power development that manage to fulfil the criteria for minimum distance to dwellings and other interests.* (Gotlands Kommun, 1999)

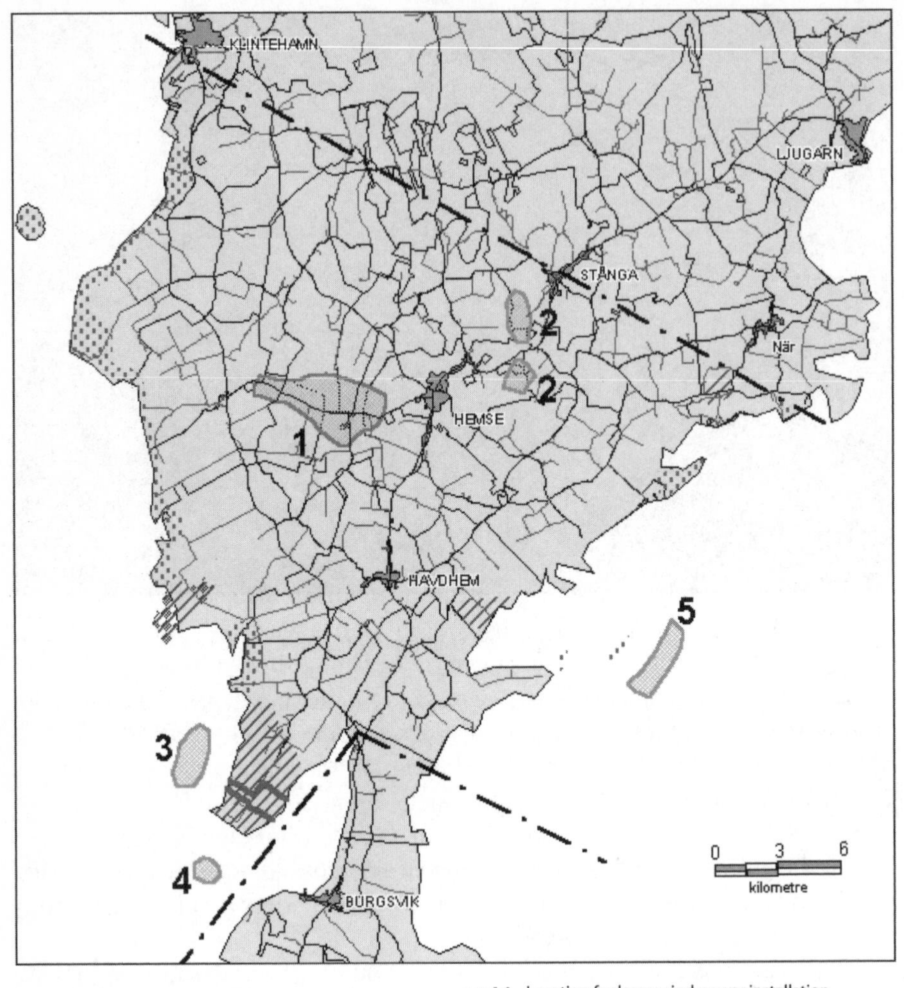

Area 1. Mästermyr *(farming landcape)*

Area 2. Bog *(farming landscape)*

Area 3. Klasården *(offshore)*

Area 4. Bockstigen *(expansion of offshore wind farm)*

Area 5. Outside Ytterholme, Grötlingboudd *(offshore)*

Location for large wind power installation (sanctioned by municipality in 1988)

Extension of area for location of large wind power installation (sanctioned by municipality in 1988)

Area where wind turbines should not be installed. The same restrictions apply also for the water outside these areas.

Development area for wind power

Area on Näsudden intended for technical development of wind turbines. Within this area deviation from regulations concerning hub height close to the coast, i.e. maximum 60 metres, is permitted. Within the area the developer in the application for building permission has to show how the development is adapted to the 'pattern' for the whole area.

Figure 14.6 *Wind power plan for southern Gotland*

The complete plan consists of a 50-page document that gives background and detailed descriptions of the five areas designated for large wind power development, including photomontages that show how these areas will look when wind turbines have been installed. Näsudden and other areas that the municipality approved back in 1988 are also included. The plan has been worked out in close dialogue with local inhabitants, wind power developers and other interested parties. The plan also has strong political support. Criteria for distance to dwellings are also included in the plan: the basic rule is 1000 metres, but the building committee may consider a protection zone of 500–1000 metres.

This means that the minimum distance to dwellings should be 500 metres, with preferably 1000 metres. Another rule says that the distance between new installed turbines or wind farms and existing wind turbines should be at least 3 kilometres. This does not, however, apply to turbines that are installed next to a farm or an already existing turbine (to extend it to a wind farm).

The plan designates both areas for large wind power installations and areas where wind turbines are *not* allowed. In other areas applications for single turbines or small groups of turbines will be examined as before the plan was made, according to the building and environment laws. The plan does not guarantee that applications for building permissions will be approved within the designated areas, since these kind of plans (comprehensive municipal plans) do not have any legal status.

After this plan was adopted a proposal for wind power development areas has been also worked out for the rest of the island. This proposal was ready in 2002, but had still not been adopted in 2006. Since work on the new plan has been ongoing, no applications for new wind power installations have been processed. The planning work itself has proved to be an efficient barrier to wind power development on Gotland.

An evaluation of the plan for southern Gotland is not very encouraging either. Since the plan was adopted, no new turbines have been developed within the new areas. There is a large offshore project in the pipeline for area 3, but no decision has yet been made to actually build it. In the area north of this, an extension of an area from 1988, a large onshore wind farm was approved by the municipality, county and the environment court, but turned down by the government. In area 1 project development started but has now been abandoned due to difficulties to connect it to the grid. Area 2 has too low wind resources, and no developer has shown any interest there; nor has any interest been shown in developing area 5 (offshore), due to the lack of grid capacity.

Regional planning

In Germany, with some 16,000 wind turbines online, spatial planning for wind power has been conducted at a regional level. One example of a regional plan and how it has been worked out will be presented here. This description is based on a paper by Matthias Plehn (2005) from the regional planning office in Rostock.

The Mittleres Mecklenburg/Rostock Region is one of four planning regions in the federal state of Mecklenburg-Western Pomerania in the northeast of Germany (see Figure 14.7).

By the end of the 1990s there were around 70 wind turbines erected in the Rostock region. The project developers chose the locations. The regional planning

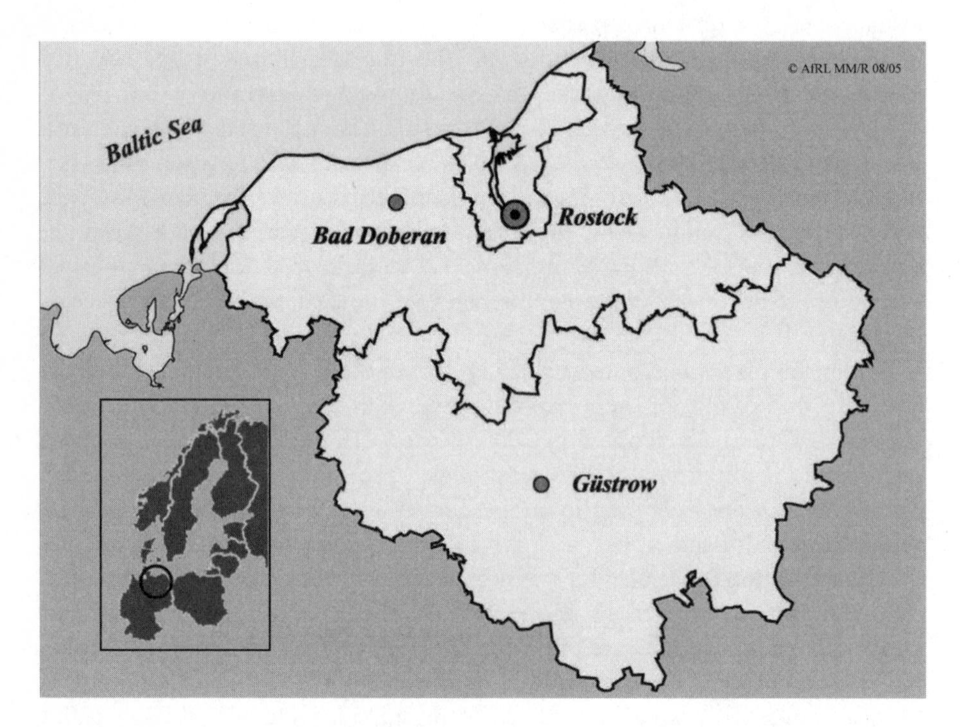

Figure 14.7 *Rostock planning region*

The planning region consists of the city of Rostock and the counties of Bad Doberan and Güstrow, with 128 communities. The planning association is responsible for the regional development programme, which sets the framework for local land use planning and for giving permission to install wind power plants. The region has 430,000 inhabitants, around 200,000 of them in the city of Rostock, the economic and industrial centre of Mecklenburg-Western Pomerania. Agriculture and tourism are important bases of the economy. Around 70 per cent of the region is agricultural land, 15 per cent forest and 10 per cent settlements and roads. One third of the region's area consists of legally protected nature reserves, landscapes and bird protection areas. The coastline extends for 70 kilometres.

Source: Plehn (2005)

authority evaluated projects individually and concerned authorities and municipalities were given the chance to make their statements about the project at an early stage of planning, so that problems could be identified before the actual permission procedure, which included checks of building regulations, nature conservation and noise prevention requirements.

Single turbines and small wind farms were installed in all parts of the region, at near-shore locations as well as in the interior. The average power was 430kW and the average plant was of only two turbines, with the largest wind farm consisting of six.

In northern Germany wind turbines had begun to change the scenery and there was an obvious need for better control by planning authorities. Changes were made in German town and regional planning law in 1997 to strengthen the legal instruments pertaining to planning outside built-up areas.

Suitable areas

According to German town planning law, buildings are generally not allowed in undesignated areas outside settlements and development sites. Only farm buildings, public supply facilities and certain projects that have a severe impact on the environment are 'privileged' to be built in outlying areas. In 1996 wind turbines were added to the list of so-called privileged projects. In the following year *concentration zones* were introduced into German town and regional planning law to avoid uncontrolled development of wind power plants. Regions and municipalities may designate certain parts of their territory for wind turbines and thereupon prohibit wind power developments in the rest of the territory.

For wind energy plants, *suitable areas* were defined. In the state of Mecklenburg-Western Pomerania the designation of suitable areas was undertaken at the regional planning level, using common criteria for all planning regions. Preliminary investigations were carried out by order of the state government of Mecklenburg-Western Pomerania. The formal designation of wind energy areas had to be finalized by the four regional planning associations which are responsible for setting up the regional development programmes.

Suitability analysis

The preliminary investigations in Mecklenburg-Western Pomerania started in 1992–1993 and were finished in 1997. The investigation was carried out in five parts:

1 **Landscape quality (visual appearance)**
The whole state territory was divided into small units, bounded by visual barriers like edges of forests, hills or settlements. These units were assessed and classified by experts. Assessment criteria were the visible diversity,

naturalness, beauty and uniqueness of the scenery. Landscape units of which the visual quality was estimated as 'medium', 'high' or 'very high', were excluded. These areas cover more than 50 per cent of the state territory.

2 **Important biotopes for wild animals**
The basis of the assessment was the current type of land use and vegetation. All available data about the occurrence and distribution of wildlife species in the state was taken into account to further differentiate area evaluation. The result was four different categories. Areas of 'medium', 'high' or 'very high' importance for wild animals were excluded from further planning. Areas of these three categories amount to 50 per cent of the state territory, much more than the legally protected nature reserves. Furthermore, around these areas, additional buffer zones were defined.
3 **Migratory bird density**
The shoreline of the Baltic Sea, rivers and lakes serve as guidelines for migratory birds. Lakes and adjacent fields are important resting and feeding places. Three different zones of bird migration density were defined. The areas of high density were excluded from further planning. These areas cover the whole Baltic Sea shoreline and the main rivers and lake systems.
4 **Wind speed**
Average wind speed was calculated for 30 metres agl, using a 1000 metre grid. A wind speed of 5.2m/s was regarded as the limit of economic feasibility, but areas with a wind speed between 5.0 and 5.2m/s were taken into consideration as well.
5 **Protection zones**
Zones around human settlements, infrastructure, forests, lakes and rivers were set. The minimum setback distances were defined for wind turbines of the 1MW generation, assuming a total height of around 100 metres (see Box 14.2).

As result of the preliminary investigations – regardless yet of wind conditions – only 1.4 per cent of the state's area remained without restrictions. These areas were regarded as being available for wind energy use (see Figure 14.8)

Suitable areas for wind turbines

The Rostock regional development programme was updated in 1997/1998 to include the wind turbine areas and approximately three-quarters of the proposed areas were incorporated into the draft regional programme. After a hearing of concerned authorities and municipalities, however, the wind turbine areas diminished to a quarter of those originally proposed, due to previously undetected local obstacles pointed out by the involved authorities.

Several proposed areas that were situated very close to each other were withdrawn to avoid a concentration of concentration zones. But the main reason for

BOX 14.2 ZONING CRITERIA FOR REGIONAL PLANNING, 1997–1998

The following list defines the criteria for distances between wind turbines and settlements, infrastructure, etc. 'Minimum distance' means the total height of a wind turbine.

Farmhouses	300 metres
Villages, rural settlements	500 metres
Towns, urban areas	1000 metres
Campsites, holiday homes	1000 metres
Motorways, federal highways	overall height of wind turbine (at least 50m)
State and county roads	overall height of wind turbine (at least 50m)
Railways	overall height of wind turbine (at least 50m)
Radio relays	100 metres
Overhead electricity lines > 20 kilovolt	50 metres
Military facilities	outer protection zone
Airfields	protection zone
Nature preservation	minimum distance
Forests, tree-lined alleys	200 metres
Coastal waters, lakes > 100 hectares	1000 metres
Rivers of first category	800 metres
Small lakes (1–100 hectares)	400 metres
Important biotopes for wildlife animals	minimum distance
Areas of very high importance	1600 metres
Areas of high importance	800 metres
Areas of medium importance	200 metres

(These areas were identified by special investigation, as described above, and cover a much larger proportion of the state territory than the officially protected areas of nature conservation.)

Areas of natural beauty
Landscape units of high or very high visual quality were excluded;
Landscape units of medium visual quality to be examined individually, mostly excluded.
(These areas were identified by special investigation, as described above, and cover a much larger proportion of the state territory than the officially designated areas of landscape protection.)

Bird migration corridors	areas of high migrant bird density excluded
Wind conditions	average wind speed 30m agl < 5m/s excluded.

this reduction was that many communities refused to have wind power plants on their territory. The regional plan is legally binding for local planning, but the consent of the local council was made a condition for the designation of suitable areas.

The results of the planning process were 25 suitable areas, a total area of 1100 hectares, 0.3 per cent of the region's territory. These areas were officially

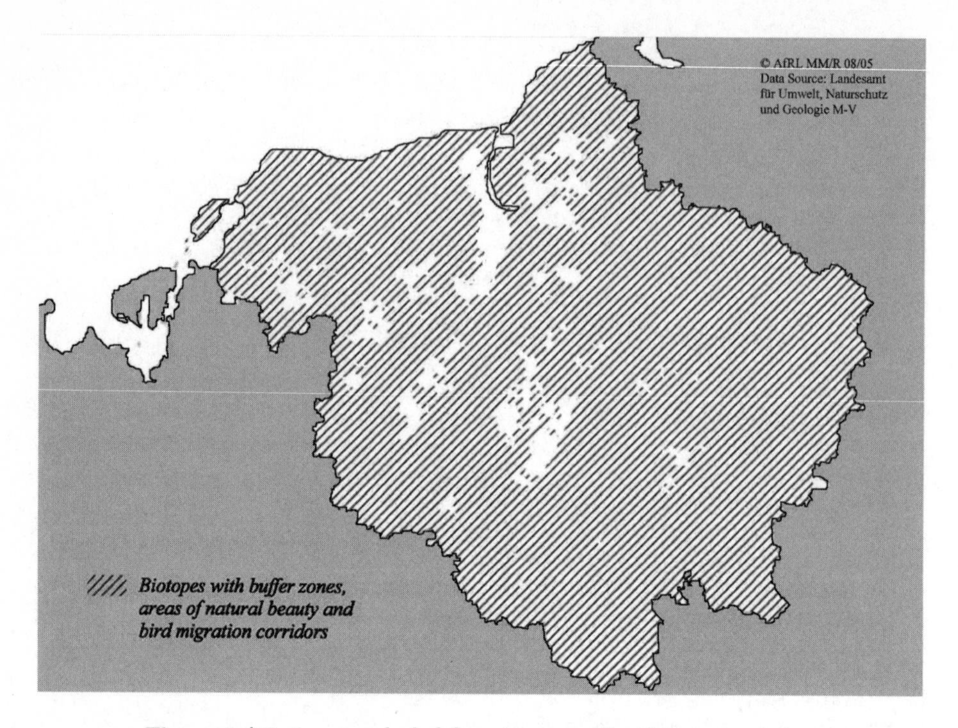

Figure 14.8 *Areas excluded for nature and landscape protection*

The near-shore areas that have the best wind conditions were completely excluded, because exclusion zones for reasons of bird migration, landscape quality or biotopes covered them all. A size of 20 hectares was regarded as minimum for a suitable area. All smaller areas were not taken into further consideration. After taking out small and low-wind areas, 1.2 per cent of the state's territory came out as usable area for wind turbines.

Source: Plehn (2005)

designated in the regional development programme, which was approved by the responsible state ministry in 1999 (see Figure 14.9).

The municipalities were encouraged to define these areas more precisely at the local planning level. Of the 25 wind turbine areas, 17 were adopted in local land use or project plans. In cases where communities did not set up land use plans, the regional development programme is the basis for permission for turbines (see Figure 14.10).

Turbine locations were often fixed by negotiation and informal agreements between the project developer, the local authorities and the regional planning authority. Changes were approved without conforming changes to the regional plan. This kind of informal adaptation and the generous attitude towards local desires has to be questioned due to the increasing willingness of project developers to take their cases to the courts and sue for permission: obvious deviations from the regional plan make it quite easy to find a cause for legal action. In future updates of the regional development programme, area outlines will be more strictly

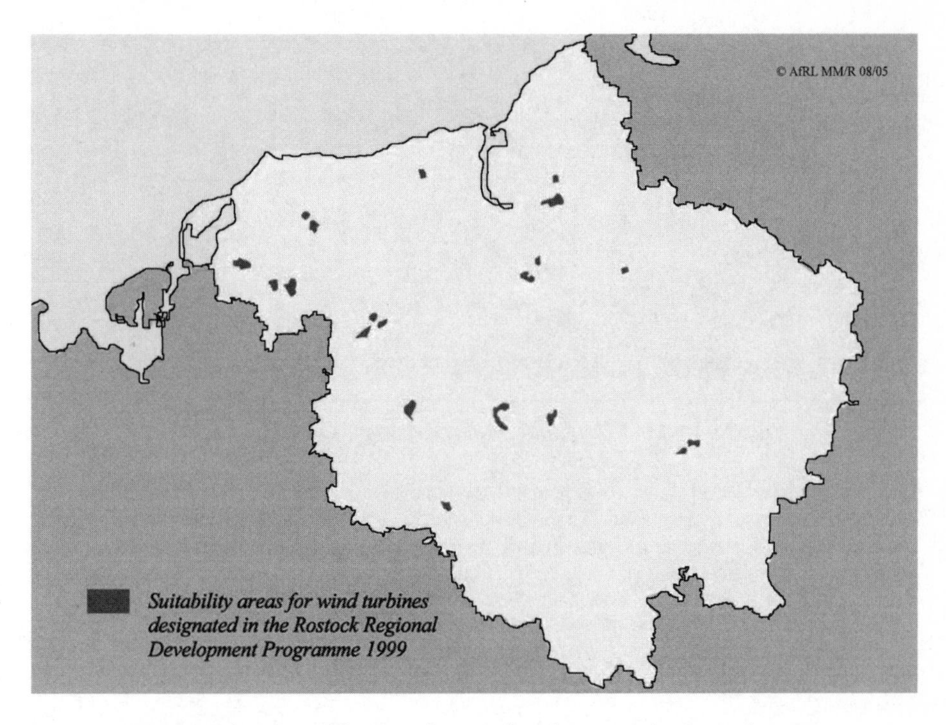

Figure 14.9 *Suitable areas for wind turbines in the Rostock region*

The results of the regional planning process were 25 suitable areas for wind power, in total 1100ha, 0.3 per cent of the area in the region.

Source: Plehn (2005)

binding. The influences of objective criteria and political will in the planning process shall be clearly separated.

Permission Procedures

Until the end of 2004 permission procedures depended on the size of the project. For small plants of one or two turbines, building permission was given by county and city administrations. Since 2005 all turbines with a total height of more than 50 metres require a licensing procedure according to the Federal Immission Control Act. The state environmental authorities are responsible. A public hearing is only mandatory for larger projects, which require a formal EIA (see Box 14.3). In the Rostock region, there had been no formal EIA of any wind power project until 2005.

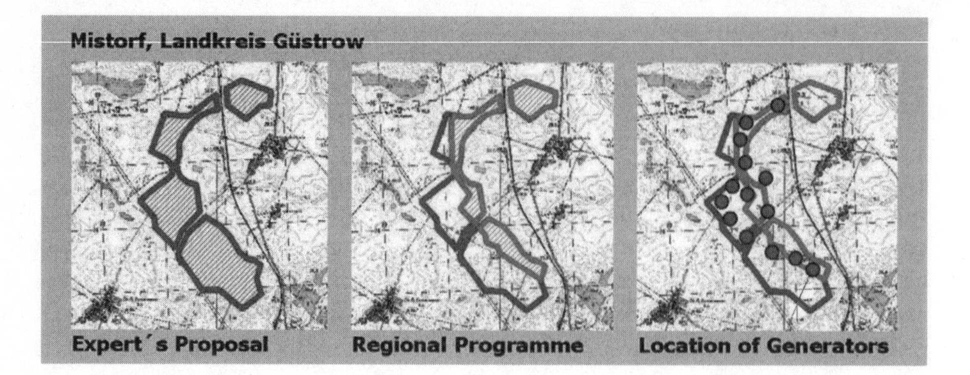

Figure 14.10 *Municipal plan in Mistorf, Gross Schwiesow*

This example from Mistorf shows an area that is divided by a municipal boundary. Since consent of local councils was set as a condition for the designation of wind turbine areas, only the eastern part of the originally proposed area became an area of suitability. Gross Schwiesow council rejected the western part. The western part is, however, more suitable for large wind turbines, because some dwellings lie close to the eastern part. Gross Schwiesow council eventually changed their minds so that the area could be used in an efficient way.

Source: Plehn (2005)

Outcome and future

By the end of 2004, 121 turbines have been erected inside the 25 suitable areas, covering a total area of 1100 hectares, which were officially designated in the Mittleres Mecklenburg/Rostock regional development programme in 1999. About 15 turbines are still projected but 85–90 per cent of the areas are already occupied. There is an average of six turbines inside each designated area and the average rated power is 1.4MW. The designation of concentration zones in the regional development programme has been successful, and no new developments have taken place outside these concentration zones (see Figure 14.11).

There is still a strong demand for development of new wind power projects and new wind turbine areas. Zoning criteria in Mecklenburg-Western Pomerania in the late 1990s were defined for wind turbines with a total height of 100 metres, but technological developments during the last five years have led to ever-larger wind turbines. Although larger turbines do not necessarily make more noise, their impact is farther reaching. Regulations for noise and shadow are no longer sufficient and the state government has issued new recommendations on minimum distances around human settlements (see Table 14.3).

If the new thresholds were applied to the existing suitable areas, most of them would have to be withdrawn. Only 20 per cent of the designated areas are situated outside the newly defined protection zones around settlements. The lower threshold values for turbines below 100 metres are met by approximately 50 per cent

BOX 14.3 ENVIRONMENT IMPACT ASSESSMENT

According to the Environmental Impact Assessment Act (2001) in Germany an EIA is an integral part of the legal permission procedures. Obligation of formal environmental impact assessment depends on the size of the project. The current (2005) rules for wind power projects are:

- For turbines < 50m no assessment is necessary.
- For wind power plants of 1 or 2 turbines, no assessment is necessary.
- For wind power plants of 3–5 turbines, assessment obligation is subject to consideration of licensing authority: assessment shall be made only if the environment of the proposed site is likely to be extraordinarily sensitive.
- For wind power plants of 6–19 turbines, assessment obligation is subject to consideration by the licensing authority.
- For wind power plants of 20 or more turbines, environmental impact assessment is mandatory.

Turbines of different owners to be built at the same location and at the same time are to be regarded as one project. If turbines are added to an existing plant, and the threshold for assessment is reached, environmental assessment is necessary.

The assessment and permission procedure has the following steps:

- first hearing of concerned authorities, definition of the scope of investigations;
- environmental investigations and documentation of expected impact by the project developer;
- consultation of authorities and participation of the public;
- consideration of hearing results and investigation outcomes by the permission authority; and
- conclusion of the expected environmental impact and decision on the project.

of the designated area. Less than half of the existing turbines in Rostock region's suitable areas could be erected again at the same location.

In Germany approximately 0.5 per cent of the territory will be available for wind power plants, according to an expert report to the German environment minister. The current proportion in the Rostock region is 0.3 per cent. Since Mecklenburg-Western Pomerania is a coastal state with good wind conditions, the proportion of wind energy should be higher than the national average, Matthias Plehn from the regional planning office in Rostock argues (Plehn, 2005).

The growing size and farther impact of turbines require a complete revision of the zoning criteria applied in the 1990s. Exclusion zones for nature conservation in the 1990s were fixed quite generously, because there was comparatively poor knowledge about the impact of wind turbines on wild animals. For the revision of regional development programmes, the exclusion zones for nature conservation purposes are to be redefined. More precise and current data, especially on migratory birds, is now available, and new protection zones will be defined around resting, feeding and

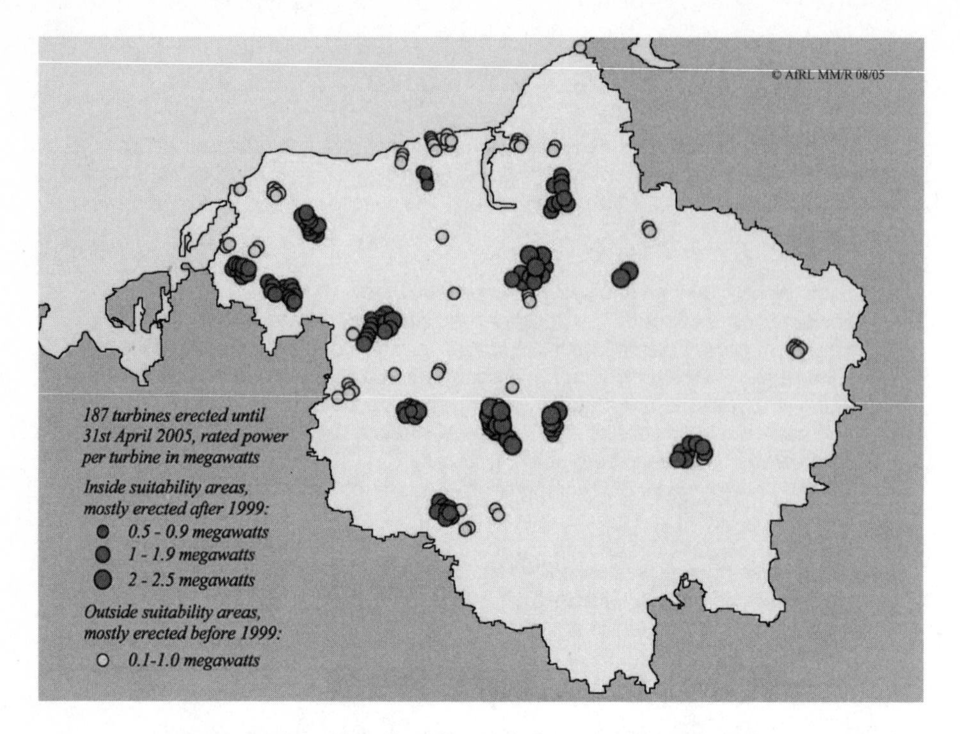

Figure 14.11 *Wind turbines in the Rostock area in 2005*

By the end of April 2005 there were 187 wind turbines online in the Rostock planning region. In Mecklenburg-Western Pomerania wind power covers almost 30 per cent of the state's electric power consumption.

Source: Plehn (2005)

Table 14.3 *Minimum distances of wind turbines to dwellings*

	1998	2004
Hamlets, farm houses	300 metres	800 metres
Mixed use areas, village areas	300 metres	1000 metres
Residential areas	500 metres	1000 metres
Residential-only areas, campsites, holiday homes	600 metres	1000 metres

Recommended distances for wind turbines, to be applied by local planning authorities, according to Mecklenburg-Western Pomerania State Government instructions from 1998 and 2004. Categories and threshold values in the first column slightly differ from the criteria in Box 14.2, which were defined in 1997. The 2004 values in the second column refer to turbines with a total height of more than 100 metres. For new turbines of less than 100 metres, the thresholds are still 500 instead of 800 and 800 instead of 1000 metres.

breeding places of birds and bats, and along their major migration routes. By such a revision space can be found for further development of wind power in the region.

Second generation planning

In Denmark and Germany, where thousands of wind turbines are online and some are approaching the end of their technical lifetimes, planning for a second generation of wind turbines has started, referred to as G2 planning. In these countries land with reasonable wind resources have become scarce. Turbines installed in the 1980s or the early 1990s already occupy the best sites. By replacing some of these small and old turbines with a few large ones, the number of turbines can be reduced while the power produced can be significantly increased.

The technical lifetime of a wind turbine is usually estimated to be 20–25 years, however, and to remove and replace turbines before they have reached this stage has proved a difficult matter. For the owner of the old turbines, there is no obvious reason to take them out of service. The investment has been repaid, so there is no capital cost left to pay and these turbines have become very profitable. The land lease contract for a good site is also a valuable asset.

In Denmark politicians introduced measures to introduce second generation turbines back in the 1990s by offering owners of old turbines a premium for dismantling them. Very few were tempted by this offer, but new regulations with the same intent have recently been introduced. According to this programme 175 new large turbines shall replace 900 small turbines, a generation shift that will increase the installed wind power in Denmark by 350MW.

This process has got under way but the generation shift is progressing more slowly than intended. This is also due to difficulties with planning. Sites for small turbines are often not suitable for large ones, since distance to dwellings criteria can't always be fulfilled: in Denmark the rule for minimum distance to dwellings is four times the hub height. The counties and municipalities have also had difficulties in finding sites for new large turbines on land. Still, many old turbines have been taken down, and for a couple of years there has been quite a large market for second-hand turbines, which after a retrofit can be installed on farms and similar sites. Many of these are also being exported to other countries.

A G2 planning project has also started at Näsudden on Gotland. In this project 17 turbines with 150 to 500kW nominal power will be replaced by six 3MW turbines. The small turbines have many different owners and some are operated by wind power cooperatives with hundreds of members; all have to agree on the deal. An application for this G2 project is under way, however, and if successful the number of turbines will be reduced by two-thirds and the installed power will increase at least fourfold, with power production increasing even more since the hub heights of the new turbines will much greater. This G2 project has received very strong support from the municipality, which expects it to reduce the visual impact on the environment.

Planning methods

In this chapter several different approaches to wind power planning have been presented, some that have proved to be efficient and others that rather act as barriers to wind power development. When targets are set, these can be interpreted as minimum or as maximum demands. The first interpretation will be a driver to wind power development, the second a barrier – when the target is reached no more wind turbines will get permission. It has to be made clear that targets are minimum demands, 'floors' and not 'ceilings'. It is also important to set these targets into a time frame.

The development of wind power planning has followed a similar path in most countries and regions. After a period without planning, the demand for planning has grown when the number of turbines has increased. This seems quite logical. An important distinction, however, is the difference between negative planning, defining areas from which wind turbines should be excluded, and positive planning, designating areas suitable for wind turbines. Negative planning acts as a barrier, positive planning as a driver. Which of these approaches is applied depends on political decisions.

Wind turbines are relatively new structures to fit into a spatial planning context. In existing regional and local plans, most areas have already been defined and in some cases also protected for a specific purpose, nature protection, infrastructure, agricultural land and so forth. On top of this, criteria for noise, shadow impact and minimum distances have to be added. When all these zones have been excluded, the areas that remain can be used for wind power. In Germany this turned out to be somewhat less than 1 per cent, which in most countries would represent quite a considerable area to be designated for wind turbines. However, when all these areas have been used, the preconditions for this kind of planning have to be reconsidered, which is what they are now doing in Mecklenburg-Western Pomerania and other regions of Germany.

It is quite easy for planners to be guided by the simple yes–no dichotomy. In the municipal wind power plan for Gotland, planners have been more nuanced. All coastal areas on Gotland are of national interest for recreation (some also for nature preservation, birds, etc.). However, in the plan some coastal areas have been designated for wind power while others have been excluded. The choice has been guided by which coastal areas are actually used by tourists and islanders for recreation and to get a view of the sea.

The borders of areas for nature protection and so forth have in many cases been quite generously set, often with additional buffer zones. But experience of the actual impacts from wind turbines on different aspects of the environment and on neighbours should be taken into account to adapt the criteria used in designation. All protected zones have a purpose; the question that should be evaluated in the planning process is whether wind turbines sited in or close to such a zone will have a negative impact on that purpose. According to Matthias Plehn from the regional planning office in Rostock, this is what planners in Germany now are doing.

There are many very good planning tools available nowadays, and most planning offices at national, regional and local levels use GIS. The most important aspect when planning for wind power is the energy content in the wind. Wind resource maps are not included as standard in ordinary GIS software, but in fact they should be the starting point for all wind power planning since small changes in average wind speed make a big change in energy content and have an even larger impact on the economic feasibility of wind power projects (see Figure 14.12). There are wind resource maps available for most countries nowadays, and these have to be included in the planning methods and procedures that are applied.

There are, however, also special planning tools for wind power planning, like the WindPLAN module in WindPRO. This is also a GIS program, but it has been developed especially for wind power applications. There are tools for checking distances to dwellings, creating maps that show zones of visual impact (where in the area the turbines can be seen) and many other useful applications. Wind resource maps can also be imported or developed. The most interesting tool in this context, however, is the one for *weighted planning*.

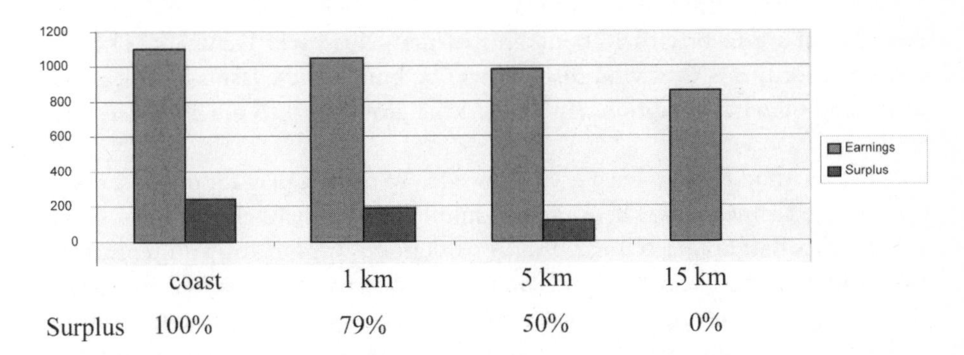

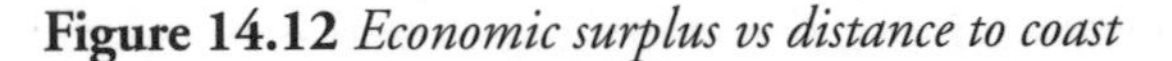

Figure 14.12 *Economic surplus vs distance to coast*

The diagram shows the relationship between economic surplus (annual earnings minus costs) and how the surplus will change when the energy content decreases with the distance from the coast for a 1MW wind turbine. The relationship between economic surplus and energy content is basically similar in all countries; quite moderate reductions of annual earnings will wipe out the entire surplus. The diagram is based on conditions in Sweden: at the coast the turbine will produce 2.2 GWh a year; annual earnings SEK0.5/kWh; SEK1.1 million/year; capital costs (6 per cent interest, 15 years depreciation) SEK782,000/year; O&M SEK80,000/year; economic surplus at the coast SEK238,000/year).

Source: Tore Wizelius

Weighted planning and the 'round-table' method

The weighted planning method is based on fuzzy logic: instead of the simple yes–no dichotomy a scale with several values (1–3, 1–5, 1–10, in fact anything but 1–2) is used. Wind resources, nature, visual impacts and so forth can be divided into, for example, five bins. On the map different areas are assigned values (for wind resources, nature, etc.) and the program will produce a map that shows where the best sites can be found after both wind resources and restrictions have been taken into account. This method also includes a subjective bias, since the values have to be assessed by someone before they are entered as input.

Weighted planning, of course, can be (and is) also used without special software. The values for different aspects in different areas can be entered in a decision table. In Mecklenburg-Western Pomerania an option would be to decide not to exclude the whole coastal zone from wind power development, but to restrict, say, only 90 per cent of it. The remaining 10 per cent could then be identified by giving different parts of the coast different values, and weight these in relation to the wind resources. A method using decision tables is also used in EIAs. This kind of planning will probably be more common in the future.

Another very efficient way of planning is the *round-table* method. This was described above in Box 14.1 about Thisted municipality in Denmark. The precondition for this is that wind turbines will be built – the question is not if, but where and on what conditions. In Thisted the group worked out a plan in three meetings.

This method has also been used in Sweden to make a plan for offshore developments in Kalmarsund. There are four municipalities that belong to two different counties that border Kalmarsund. A workgroup was formed with representatives from these municipalities and counties, civil servants and politicians. In only two years, after preparing and evaluating all the necessary background material, and after some quite heated discussions, they agreed on a common plan where areas for offshore wind farms were designated and incorporated into the comprehensive municipal plans. The *one-stop shop* permission process in the UK for offshore developments can also be described as a round-table method.

15

Opinion and Acceptance

When wind power is developed it is important that the inhabitants that live in the vicinity of the wind turbines accept these new elements in their living environment; that wind power has a high *acceptance*. In many countries it is easy to get the impression that there is strong opposition to wind power among the public when you read reports and letters to the press about protests against planned wind power projects.

If a systematic review of the press coverage is made, however, you will most likely find as many articles that are positive or at least neutral to the subject of wind power, as well as letters to the press that promote the development of this renewable energy source. In several countries there are some organizations that very actively oppose wind power, like Country Guardians in England, the Association for Protection of the Landscape in Sweden and Windkraftgegner in Germany, but these only represent a minority in these countries (see Figure 15.1).

How wind turbines are perceived is a subjective matter: different individuals perceive wind turbines in different ways. People can also change their opinion in time when they get used to wind turbines. There are, of course, some wind

Figure 15.1 *Anti-wind power associations on the internet*

Source: www.countryguardian.net

power projects that have been planned without regard for the natural environment, visual intrusion or impact on neighbours, and sometimes people who have a positive attitude to wind power in general can have good reasons to oppose a specific wind power project.

Although attitudes to wind power are subjective, it is still possible to find objective measures of public opinion in surveys carried out using proper scientific methods. And *almost all opinion polls and surveys conducted so far show that a vast majority of respondents have a positive attitude to wind power*. This applies both to national surveys and to surveys among inhabitants in areas with many wind turbines installed in the vicinity (see Box 15.1).

BOX 15.1 RESULTS FROM OPINION POLLS IN SELECTED COUNTRIES

Denmark

A nationwide survey in 1993 in Denmark showed that 82 per cent of Danes wanted more wind power and that 61 per cent thought that wind power was well suited to the landscape. The survey also showed that Danes who had a wind turbine within sight of their home, workplace or school had the most positive attitude to wind power.

Another nationwide survey, in 2001, posed the question: *Should Denmark continue to build wind turbines to increase wind power's share of electricity production?*

Yes, answered 68 per cent of respondents, while 18 per cent found the current level satisfactory, 7 per cent were of the opinion that there were already too many and 7 per cent were undecided (Danish Wind Turbine Owners' Association, 2002).

Germany

Germany is the country with the most wind power installed in the world, and also where development has been fastest. A nationwide survey in Germany in 1997 showed that 88 per cent of Germans wanted to have more wind power. Among parents with one or two children, 94.5 per cent wanted more wind power.

A similar survey conducted in 2002 showed that 88 per cent of respondents supported the construction of more wind farms in Germany as long as certain planning criteria were met. Only 9.5 per cent considered that there were already enough (Wind Directions, 2003).

The UK

Many opinion polls have been conducted in the UK by various organizations since the first wind turbines were installed in 1991. The British Wind Energy Association has made a summary of 42 different surveys carried out from 1990 to 2002. The summary shows that 77 per cent of the public are in favour of wind energy, while 9 per cent are opponents. A survey of 2600 persons in 2003 came to a similar result: 74 per cent of respondents supported the government's aims to generate 20 per cent of the UK's electricity from renewable energy sources by 2020, and further development of wind power, 7 per cent were against and 15 per cent were neutral (Wind Directions, 2003).

France

In 2003 a survey of 2090 persons was conducted in France: 92 per cent of respondents were in favour of further development of wind energy, considering both the environmental and economic advantages of the technology, and also as a substitute to other energy sources, including nuclear power (Wind Directions, 2003).

The US

According to a national survey in 2005, 87 per cent of respondents considered it a good idea to build more wind farms (Yale University, 2005).

Australia

In a nationwide survey in 2003 building new wind farms to meet Australia's rapidly increasing demand for electricity was supported by 95 per cent of respondents.

Australia has large coal mines and many coal-fired power plants, and a strong lobby that protects the coal industry. However, for 71 per cent of the respondents, reducing greenhouse pollution outweighed protecting industries that rely on reserves of fossil fuels (Wind Directions, 2003).

Sweden

The attitude of Swedes is accounted for in reports from the SOM Institute at Gothenburg University. This nationwide survey, conducted in 2000, showed that 73 per cent of respondents had a positive attitude to wind power, not only in general, but also to installations of wind turbines in their own municipality. Only 9 per cent were negative, while 18 per cent didn't express any opinion.

In another report from the SOM Institute from 2004, 64 per cent of respondents expressed the opinion that investments in wind power should increase, with 22 per cent believing that they should remain at the current level. This means that 86 per cent of Swedes want wind power to be further developed in Sweden (Hedberg, 2004 and Wizelius et al, 2005).

According to all these surveys, most of them conducted in countries with significant amounts of wind power installed, there is a very broad acceptance among the public of wind power in general, and the benefits of this renewable energy source in particular. It seems that general acceptance of wind power is high in all countries. This does not necessarily imply, however, that the respondents would accept wind turbines in their immediate environment.

The best wind resources are found at sea and along the coasts, and most wind turbines are installed close to the coast. A closer look at the reports from Sweden shows that people who live along coasts are just as positive to wind power as people who live inland. It shows that people who already have wind turbines in sight in the landscape where they live generally don't have any objections. There were no differences in attitude related to sex or education, though the very young and very old were a little less positive than the average and people from urban areas, Stockholm especially, were less positive than those living in the countryside.

BOX 15.2 OPINION ON WIND POWER, SELECTED AREAS

The UK

In the UK less than two out of ten would oppose development of wind power in the vicinity of their dwellings. More than a quarter would have a very positive attitude to this, according to a national survey from 2003. Of respondents that lived in areas where there were already wind turbines installed, 94 per cent would be positive to a further development and only 2 per cent were negative (Taylor Nelson Sofres, 2003).

Scotland

According to a survey conducted for the Scottish Executive, 82 per cent of those who lived close to Scotland's ten largest wind farms wanted more electricity generated by wind farms, and 50 per cent supported an increase of the number of turbines at their local wind farm.

The poll covered 1800 people living within three zones, up to 5km, 5–10km and 10–20km away from operating wind farms. People who lived in their homes before the wind farm was developed say that, although in advance they thought that problems might be caused by its impact on the landscape (27 per cent), traffic during construction (19 per cent) and noise during construction (15 per cent), the reality reduced these figures to 12 per cent, 6 per cent and 4 per cent receptively (Wind Directions, 2003).

The Aude region, France

In a sample of 300 persons living near wind turbines in the Aude region of southern France, 46 per cent agreed that wind turbines affect the countryside, but 55 per cent considered wind farms to be aesthetically pleasing (Wind Directions, 2003).

Cape Cod, US

A planned offshore wind farm in the Nantucket Sound off Cape Cod in Massachusetts has caused a lot of debate. Opinion Dynamics Corporation carried out a survey of 600 local residents. To the question of which energy resource, coal, oil, natural gas, nuclear or wind, the respondents preferred, wind power was chosen by 42 per cent of respondents. On the question of whether they favoured the planned offshore project, 55 per cent were in favour, 35 per cent opposed it and 10 per cent were undecided (Wind Directions, 2003).

Sottunga, Åland, Finland

In a survey on Sottunga on the island of Åland in the Baltic Sea, where a wind turbine has been online for some years and more turbines were planned, none of the 55 permanent inhabitants were negative to wind power. Half of them had gained a more positive attitude to wind power since the wind turbine was installed than they had had before.

Source: Wizelius, 1993

A Danish report has shown that those who have wind turbines within sight of their dwellings, schools or places of work in fact have a more positive attitude than other people. A closer analysis shows that those who live permanently in an area with wind turbines are more positive to wind power than those who visit it or have holiday cottages there. Finally the positive attitude is stronger where people are offered the chance to buy shares in the turbines and when they have been informed about the advantages for the environment.

Many opinion polls and surveys have been made in municipalities and regions where there are many wind turbines installed and where people have practical experiences of wind power (see Box 15.2).

Stability of opinion over time

In a region where wind power has developed rapidly, one would expect that acceptance among the population would change during the years as the number of turbines increased. In some regions of Spain such rapid development started in the late 1990s, and in 2004 Spain had the fastest growth of new wind power capacity in the world. Social acceptance of wind power has been studied in three regions in Spain, Navarre, Tarragona and Albacete, where many wind turbines are installed.

In Tarragona four studies were conducted from 2001 to 2003, with 600 people in each poll. These four polls show that the strongest support comes from people living near wind farms. In Albacete a study was conducted in 2002; it showed that 79 per cent of respondents considered wind energy to be beneficial.

In Navarre a study from 2001 showed that 85 per cent were in favour of the implementation of wind power in Navarre, with only 1 per cent against. The study also showed that acceptance increases while new wind farms are being developed and installed: for most people the benefits of wind energy compensated for any negative impacts experienced during implementation (see Table 15.1).

Table 15.1 *Public acceptance in Navarre*

	1995	1996	1998	2001
Turbines*	6	72	187	659
Positive %	85	81	81	85
Negative %	1	2	3	1
Indifferent/don't know %	14	17	16	14

* most 660kW

From 1995 to 2001 the number of turbines increased from 6 to 659, while the share of the inhabitants that have a positive attitude to wind power has remained constant.

Source: EWEA (2003)

Wind turbines in the living environment

In 2004 Gotland University made case studies in three different areas on the island of Gotland in the Baltic Sea, where the inhabitants were living close to wind turbines. In the village of När everyone who lived within 1100 metres of two large wind turbines was interviewed, in Klintehamn a sample of those who could get shadow flicker during sunset, and in Näsudden those households situated in the middle of a large wind farm with 81 turbines. In total 94 persons from 69 households were included in the investigation.

Considering that all respondents lived close to wind turbines, the nuisances reported were surprisingly small. Very few of them were annoyed by noise or shadows or considered that their view of the surrounding landscape had been destroyed. Of the total number of persons interviewed, 85 per cent were *not* annoyed by noise from the wind turbines around their homes. For rotating shadows the share of those *not* annoyed was even bigger, 94 per cent. Relatively few of those living in Näsudden, where there are 81 wind turbines online, think that their view of the surrounding landscape has been negatively affected by this – 13 per cent. Of all the residents in the three areas, 89 per cent expressed the opinion that wind turbines had not disturbed their view. The acceptance of wind power among people living with wind turbines as close neighbours was high (Widing et al, 2005).

Wind power and tourism

Since wind turbines should be installed at sites with good wind conditions, many wind farms are sited at the coast, onshore and, in recent years, also offshore. Many coastal areas are also popular tourist resorts, so there is an obvious risk of a conflict of interest. The same conflicts can also occur in mountain areas, at the wind farms in Scotland, for example, and around ski resorts. Several surveys have been made to find out if wind turbines will scare away tourists from attractive holiday areas.

A survey among tourists in Germany in 2003 showed that 76 per cent considered that nuclear and coal-fired power plants spoiled the landscape, while only 27 per cent though the same about wind turbines (Wind Directions, 2003).

A survey on tourism in Schleswig Holstein showed that the wind industry does not affect tourism in the region. Visitors are aware of the increasing number of turbines, but this does not influence their behaviour (EWEA, 2003).

In Belgium there are plans to develop an offshore wind farm 6km off a stretch of coast where there are many holiday resorts. A survey conducted in 2002 by the West Flemish Economic Study Office showed that 78 per cent of the public were either very positive or neutral about this offshore wind farm (see Table 15.2).

Two separate polls have looked at the impact on tourism in Scotland. A MORI poll in 2002 found that over 90 per cent of visitors would return to Scotland for a holiday whether or not there were wind farms in the area. Another survey by the

Table 15.2 *Belgian public perception of wind farm 6km off the coast*

Group	Negative	Neutral to positive
Residents	31.3	66.5
Second residence	10.2	88.8
Frequent tourists	18.7	81.3
Occasional tourists	19.5	80.5
Hotels etc. with sea view	19.5	80.5
Other	15.3	84.7
Average	20.7	79.3

Visit Scotland tourism agency found that 75 per cent of visitors were either positive or neutral towards wind farm development in general, but less positive about their visual impact. However, those who actually had seen a wind farm during their visit were more positive than those who had not (EWEA, 2003).

There are other surveys that have investigated how different groups of inhabitants evaluate the view of the landscape. They show that permanent residents in the countryside consider the landscape as a natural resource that should be utilized in a sensible way, while those who use the landscape for recreation have a more aesthetic view and consider it as a 'picture postcard' that should remain unchanged. According to the surveys made in Scotland and Schleswig Holstein, it seems that wind turbines can be an accepted element in the tourists' picture postcards.

NIMBY attitudes

People who have a positive attitude to wind power in general sometimes can have another attitude to the development of wind power close to their homes or holiday cottages. This phenomenon is often called NIMBY – 'not in my backyard'. The Dutch researcher Maarten Wolsink, who has studied the NIMBY phenomenon, defines four different types of NIMBY reactions (see Box 15.3).

This NIMBY effect can be measured. A typical NIMBY curve starts at a rather high level. In a survey on the general attitude to wind power within a group of people, say 75 per cent express a positive attitude. When a project to install wind turbines in the vicinity starts, some of these people get worried about the impact on their living environment – noise, shadows, changed views and so forth – and the share of those with a positive attitude tends to decrease to around 60 per cent. When the wind turbines have been installed and have been running for a couple of months, the share of those with a positive attitude recovers to the initial value, actually often even higher.

The general public has a very positive attitude to wind power, according to surveys from many different countries, regions and local areas. If those that are

Box 15.3 Four NIMBY attitudes

NIMBY A

Positive attitude to wind power installations in general, but negative attitude to installations in the immediate vicinity.

NIMBY B

Generally negative attitude to wind power.

NIMBY C

Positive attitude to plans to develop wind power, which change to negative when there are plans to install wind turbines in the vicinity.

NIMBY D

Negative attitude to the planning procedure rather than to wind power.

Source: Hellström (1998)

concerned are informed about the environmental advantages – notably that wind turbines produce electric power without any hazardous emissions – the attitudes tend to be even more positive. The attitudes also depend on how people in the vicinity of a planned wind power project are informed about the plans.

The Swedish utility Vattenfall has made great efforts to inform the public about wind power, first in Lysekil before a prototype wind turbine was installed there, and then, in the form of a public consultancy, in the Kalmarsund area for a planned offshore wind farm. Several thousand households were informed by mail and a survey was conducted to get good information about the opinions of people living in the area. The project had its own website, where questions could be asked about the project.

The developers of the Utgrunden offshore wind farm in the same region also invested a lot in informing the public in Torsås municipality, which was the base for the development. Schools and local inhabitants were invited to visit the wind farm by boat while the turbines were installed and an exhibition hall with binoculars to look at the wind farm and with a wind tunnel was established in the harbour. They even printed a wind turbine colouring-book that was distributed to day-care centres and schools (see Figure 15.2). Follow-up studies have shown that these efforts have increased acceptance of wind power in the community to a considerable degree.

Some other large offshore projects in Sweden have, however, been received by strong protests. These reactions could be categorized as NIMBY D – they have not been opposed to wind power in general, but to the planning procedure and the design and size of the projects. The public information about these

Figure 15.2 *Wind power promotion material for children*

Did you Ever Hug a Wind Turbine? is the title of this booklet, with cartoons that explain some basic facts about wind turbines and the environment for the very young. The booklet was distributed to all children in day care centres, kindergartens and junior schools in the Torsås municipality in Sweden while a large offshore wind farm was being developed.

Source: Enron Wind

projects was badly managed. By providing good information NIMBY effects can be avoided.

The NIMBY concept has been under discussion for some time, and there is no doubt that a NIMBY reaction can occur among inhabitants in areas where new wind power developments are planned. However, if wind power development in a country is slow, and the targets set are not met, as in the Netherlands and Sweden for example, this cannot be blamed on resistance from the local inhabitants. The reasons are more likely to be institutional barriers: government policy and the laws and rules that regulate the permission process and the economic conditions for wind power. Institutional factors have a greater impact on wind energy facility siting than public support, according to researcher Maarten Wolsink (Wolsink, 2000).

Schleswig-Holstein in northern Germany had 1800MW of wind power installed in 2002, which produced 30 per cent of the region's energy consumption.

A study (Eggersglüss, 2002) showed that most people accept the siting of wind turbines if the following principles are followed:

- sufficient distance to residential areas;
- quiet turbines are chosen;
- the population is kept properly informed;
- there is a financial benefit for the community;
- the developer is from the area; and
- landowners' views are sought when the site is chosen.

If all wind power project developers followed these simple rules of conduct, acceptance at the local level would probably be as high as the general acceptance of wind power.

Acceptance problems at higher levels

Surveys that have been conducted to find out the public opinion on wind power prove that acceptance of wind power by the general public is not a problem. Wind power has in fact very strong support from the public. Problems with acceptance are usually found at higher levels, although this obviously differs in different countries.

In Sweden and many other countries, the authorities, politicians, grid operators, power companies and industry all have problems with accepting wind power and raise barriers that delay or stop its development. Acceptance is also low among scientists working in the energy field, since many have strong interests in conventional power technology such as fossil fuel combustion and nuclear power. Many leading scientists are still repeating the same unfounded arguments against wind power that power companies have used since development started in the 1970s.

In the article 'Phase out of nuclear power proof of ignorance' published in a major Swedish morning paper, four members of the Royal Swedish Academy of Science included the following statements about wind power:

> *The contribution to electric power production from wind power is moderate (0.3 per cent). Further development is possible and within 20 years, with massive investments, up to 10TWh (7 per cent) could be produced. Even after being fully developed, wind power would only give a marginal contribution.*
>
> *Since wind power cannot be regulated (due to variations in wind speed), power balancing is always necessary, and this can be done only by hydropower. For such a high share of wind power as 10TWh an extensive development of hydropower will be necessary as well. To that must be added that the Nordic power grid needs to be reinforced. Wind farms*

> *also need very large areas. Altogether, the cost will be significantly higher than for other power plants, which makes it impossible for wind power to compete without large subsidies.* (Kullander et al, 2002)

The type of *disinformation* about wind power that the quoted article exemplifies has been spread since wind power development started in the early 1980s. These misconceptions have been propagated by representatives of power companies and by researchers with interests in nuclear power.

In the first paragraph the authors assert that wind power can only make a marginal contribution to power production even when it is fully developed. However, the contribution of wind power obviously depends on how much wind power is installed. In Germany and Spain wind power that can produce 10TWh/year is currently developed in about two years (in 2005 Germany installed 1808MW and Spain 1764MW). With the same pace in Sweden, wind power could produce 30TWh within ten years and contribute 20 per cent of the electric power. Wind power does *not* have to play a marginal role in power production.

In the second paragraph the authors assert that an extensive development of hydropower is necessary to integrate wind power in the power system. This claim is also wrong. Grid operators had to learn to regulate the power in the power system long before wind power was at hand. The diurnal power fluctuations are far greater than could be caused by a large share of wind power in the system. The capacity of hydropower to regulate the power will actually increase when wind power is developed, since water can be saved in the hydropower dams when the winds are strong. This saved water can then be used for power regulation.

In the same paragraph they assert that the Nordic power grid needs reinforcements. But whether this will be necessary depends on the geographical distribution of large wind farms. Like all new power plants, wind farms have to be connected to the power grid. The grid itself does not, however, need any reinforcements, at least not until wind power penetration (share of total production in the power system) reaches a level of more than 10 per cent. The statements that wind power needs more space, and that wind power never can become competitive without subsidies are wrong too (see Figures 12.4, 12.5, 13.1 and 13.2).

Although these arguments against wind power have been refuted time and time again in scientific articles, government reports and newspaper articles, they reappear and are presented as undisputable facts by persons with high status in the science community. These arguments will, however, never become valid, not even by eternal repetition.

16

Grid Connection of Wind Turbines

Most wind turbines are connected to the electric power grid. Wind turbines are not only a new kind of power plant that transform wind energy into electricity, they also have other traits that power companies, utilities and grid operators are not used to. Wind speed is constantly changing, in a way that is hard to predict, and so the power production will vary. Wind turbines are comparatively small and are usually connected to the distribution grid, while large conventional power plants are connected to the transmission grid, with much higher voltage levels.

The power grid

Electricity is not an energy source, but a form of energy that is used to transport energy in the form of electric power from power plants to power consumers. The electric power grid constitutes a practical and cheap means to transport power from power plants that exploit the kinetic energy in running water (hydropower stations) or the wind (wind turbines) or that boil water (with uranium, oil, biomass or other fuels) to run a steam turbine that runs a generator that in turn produces electric power. This power is then transported on the grid to consumers, who with different types of equipment transform the electric power into light, heat or mechanical work.

When the use of electric power started around a hundred years ago, the power grid was built just in the close vicinity of the power plants. The grid at that time consisted of small isolated 'islands' in cities and around hydropower stations where there were different kinds of factories, which had exploited the energy in the running water by water wheels to run sawmills, smithies and so forth in earlier days.

Such local grids were very vulnerable. If the river ran dry or the power plant had a failure or needed servicing, no electric power could be produced. To avoid this, back-up power plants were needed. After some decades these isolated grid-islands were connected to each other to form larger regional and finally national power grids. The need for back-up decreased and was almost eliminated, since the many power plants connected to the grid could replace each other. Today national power grids are also interconnected across borders.

In most countries the backbone in the power system consists of a *national transmission grid*, with a voltage level of 400 or 230kV, that is managed by a transmission system operator (TSO). In Sweden, for example, this is the state-owned company Svenska Kraftnät AB. This very high voltage transmission grid is used to transmit large amounts of power over long distances. The next level is the *regional grid*, with a voltage level of 130 or 70kV (in Sweden), which transmits power from the national transmission grid to the local *distribution grid*. These have lower voltage levels, 40, 20 or 10kV, and are used to distribute power to factories, households and other consumers. Before the power enters the consumers' low voltage grid, the power is transformed to 690 or 400V (see Figure 16.1).

Just as the road network is used for transport of people and commodities, so the power grid is used for the transport of energy. And just as motorways are built for a large flow of heavy transports, so the national transmission grid is set up for large flows of energy, while the distribution lines in the outskirts of the grid correspond to dirt roads that are too weak for heavy lorries. Like roads for vehicles, the dimensions and voltage levels of the grid sets limits to how much power can be fed into the grid and what size of power plants can be connected to different parts of it.

In Sweden the power system is based on a few very large power plants, hydropower plants in the north of Sweden and nuclear reactors in the south. There are also some CHP plants that produce heat and electric power simultaneously and some gas turbines that are used for peak power when the loads are extremely high during cold winter days or to balance sudden increases in power consumption.

Electric power consumption in Sweden is about 150TWh/year. During different periods of the year power is also imported or exported. The share of wind power was still less than 1 per cent in 2005, compared to around 20 per cent in Denmark.

Power consumption is constantly changing, during each day and also during the year. The need for power is of course less at night, when factories are idle and people are sleeping. In the morning, when people prepare breakfast, power consumption increases fast, and you get the same increase in the evening when it is time for dinner.

When the football world championships are shown on TV, power consumptions peaks during the half-time break, when hundreds of thousand of kettles are turned on to make tea and coffee (see Figure 16.2).

In Sweden, where many homes use electric heating, electric power consumption is much higher in the winter than in the summer. The peak usually occurs in January or February, when it is cold all over the country and the wind is blowing as well. Then the power demand is at its highest and all power plants run at full capacity, with some power also imported (see Figure 16.3). In other countries the annual variations in power demand look different. In California, for example, the peak appears in the summer, when air-conditioning is running day and night.

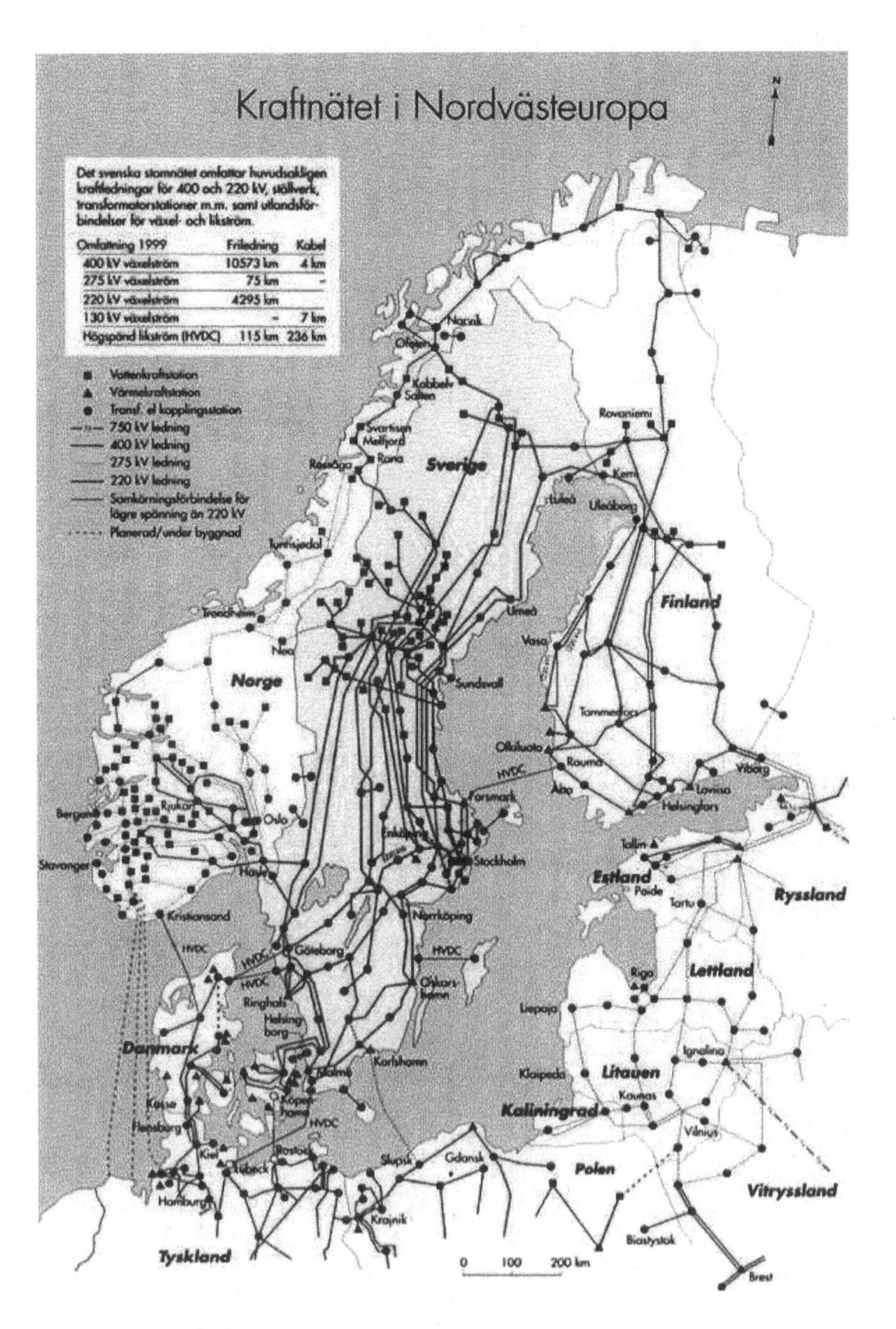

Figure 16.1 *The electric power grid in northern Europe*

The power systems in the Nordic countries, Sweden, Denmark, Norway and Finland, are interconnected (Nordel) and have a common power market (Nordpool). There are also links to power systems in other countries, for import or export of electric power.

Source: Svenska Kraftnät

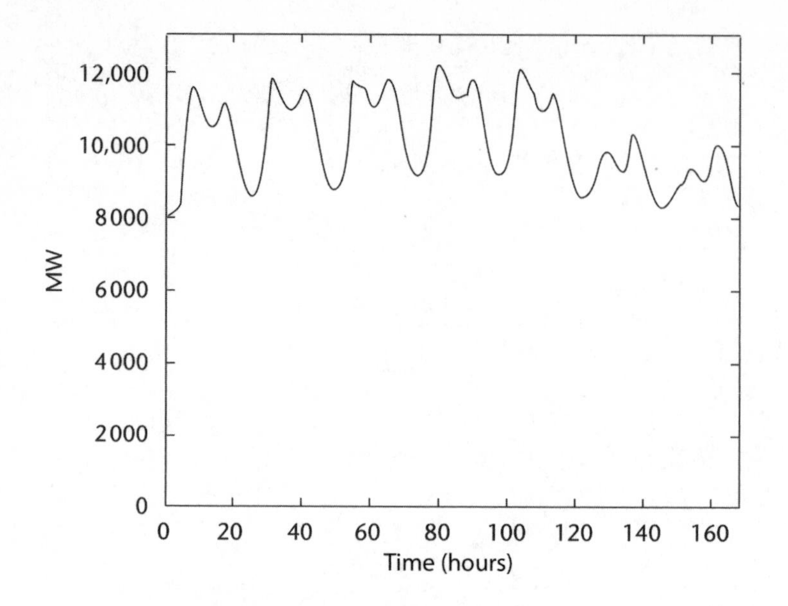

Figure 16.2 *Variations in power consumption, diurnal and weekly*

The diagram shows the variations in power consumption during an ordinary week. The vertical axis shows the power demand (power consumption) and the horizontal axis the hours during a week. Power consumption has a minimum around midnight; there is a peak around 8–9 in the morning and another one around 6–7 in the evening. During weekends the peaks are considerably lower than on weekdays.

Source: Söder (1997)

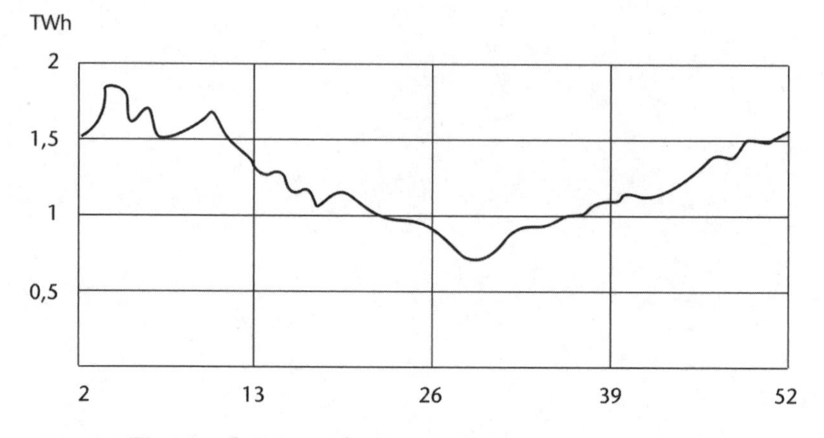

Figure 16.3 *Annual variations in power consumption*

The diagram shows the weekly variations of power consumption in a region in Sweden during one year. In a cold week in winter power consumption is more than twice as high as during the summer holiday weeks.

Source: Blomqvist (1997)

Electric power is a perishable commodity, and power production in the grid has to match the power consumption at all times. If it doesn't, the frequency will change. If power production is larger than consumption, the frequency will increase; if consumption is larger than production, the frequency will decrease. When there is a mismatch, electric equipment can be damaged or the grid breaks down and a power cut results. This very seldom happens, however, because grid operators understand how to manage the power system and to handle the balance of supply and demand.

Different power plants have different roles in the system. In Sweden nuclear reactors are used for base load; they run day and night most of the year, and it's a slow and complicated matter to regulate the power output. Hydropower is very easy to regulate, and is thus used for regulating purposes. Gas turbines can be started within seconds and are used for peak power and when there is a need for very fast power regulation. Nowadays peak loads are often handled by import of power from neighbouring countries.

Different parts of the power grid have different voltage levels. In Sweden the national transmission grid has a voltage of 400,000 or 230,000V. With high voltage, losses are reduced. A power line of any given dimension (cross-section area) can transmit more power the higher the voltage level is. To transmit large amounts of power over long distances it is most cost-efficient to use as high a voltage as possible. With all transmission of electric power, however, losses (in the form of heat) are unavoidable. How large these losses will be depends on the dimensions and lengths of the power lines. In the Swedish power system around 10 per cent of the power that is fed into the grid will be lost. The TSO is actually also the largest power customer in Sweden, since the power that is lost has to be bought and paid for.

The voltage level is reduced in several steps on the way from the power plant to the consumer. In the first step it is transformed from the national grid (230 or 400kV) to the regional grid (70 or 130 kV), then to the distribution grid (10–40kV). Before the power is fed to the consumers, it will be transformed to 690V (factories) or 400V (households). When the voltage is decreased losses increase, but power lines and other equipment will be much cheaper.

The TSO has the overall responsibility for the operation of the power system. The local grid operators, who manage the distribution grid, have to predict and order how much power they will need in the coming year, and then also specify the power needed 24 hours in advance. If the actual power consumption is higher than that, the grid operator has to pay penalties. If power production does not balance consumption, the TSO can order power plants to increase or decrease their production (see Box 16.1).

BOX 16.1 POWER REGULATION

The principle for the power system is that production with low operating costs is run all the time and the production with high operating costs is run only when the load is high. Wind power has low operating costs, and since it is difficult to regulate it has the right to feed all power produced into the grid at all times.

In a deregulated market, the task of balancing supply and demand falls to balance responsible players (BRPs), power trading companies. All production and consumption goes through a contract with someone who has a contract with a BRP.

These BRPs give production schedules to the TSO one day ahead, but they can change the schedule up to the hour of delivery. During the operating hour, responsibility shifts to the TSO.

If the scheduled power production (supply) does not fit the actual power demand, the power production has to be regulated (increased or decreased). To handle this need there is a *regulating power market*. Producers bid the regulating power to this market (offer a price for supply of additional power) and the TSO determines the use of reserves (and the regulating power market) according to the net imbalances in the Nordic system. After each operating hour the net imbalances of all BRPs are calculated and charged for according to country-specific rules.

Wind turbines in the power system

Wind turbines are quite small power plants that should be sited where winds are strong. Therefore they are often located at the periphery of the grid, where the grid is weak. One or a few turbines can always be connected directly to the distribution grid.

A wind turbine's production varies with the wind speed. Wind energy can't be used for base loads, since the production varies. Nor can it be used as regulating power, since the energy from wind turbines cannot be increased with demand. Nor can it be used as peak power, for the same reason. From point of view of the electric power system, wind power works in a completely different way from conventional power plants. Wind turbines are usually connected to the distribution grid, the same power lines that consumers (households, industries) use. They function as decentralized power plants.

The grid operator can handle the variations of power produced by wind turbines in the same way as variations in load (power consumption). When the wind speed decreases the power from other power plants, for example hydropower, will be increased. As long as the share of wind power is below 10 per cent of the total power production this is no problem (see Figure 16.4).

Power companies and grid operators are still not used to this (at least not in countries and regions where wind power development still hasn't started), but it

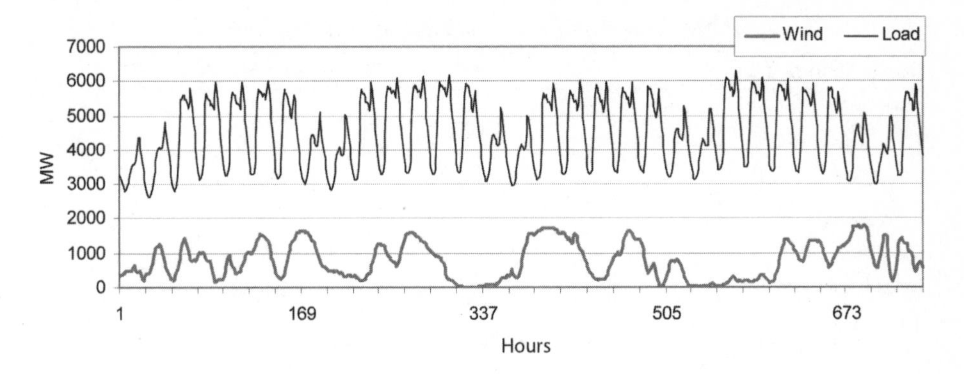

Figure 16.4 *Wind power variations*

The diagram shows how loads (consumption) and wind power production varied in January 2000 in Denmark. The variations are handled just like ordinary load variations, when wind turbines produce a lot of power the contribution from other power plants is reduced, and vice versa.

Source: Holttinen (2004)

is not a problem to integrate quite a large amount of wind power into the power system.

Grid connection of wind turbines

Wind turbines are connected to the grid via a transformer that raises the voltage from 400V (turbines up to 250kW) or 690V (larger turbines) to the grid voltage, which can be 10, 20 or 40kV (kV denotes kilovolt, 1000V; voltage levels can vary from country to country). The developer has to install and pay for the transformer, which is usually sited on the ground next to the turbine. Several small turbines can be connected to the same transformer. Larger turbines (500kW and more) have their own transformer, and MW turbines often have the transformer integrated in the tower, with the transformer included in the price of the turbine.

The distance from a wind power plant to the existing grid is an important factor to take into consideration. The cable from the turbines to the grid connection point has to be paid for by the wind power developer, and since the price per kilometre of cable is significant, the distance will have a considerable impact on the investment cost.

The capacity of the grid sets a practical/technical limit on the amount of wind power that can be connected to a specific power line. There are several technical factors to take into consideration (the dimensions of the lines, voltage level, power flows, distance to the closest transformer station, loads, etc.), and only electric power engineers can make these kinds of calculations.

There are, however, some rules of thumb that give an idea of how many MW of wind power can be connected to power lines with different voltage levels. One

Table 16.1 *Wind power capacity and voltage level*

Connection point	Maximum capacity (MW)
10kV line	1–2
10kV transformer	8–10
20kV line	5–8
20kV transformer	15–18
40kV line	13–18
40kV transformer	30–38
130kV line	30–60

Close to the transformer station, where the 10kV line is transformed to a higher voltage level, more wind power can be connected than close to the end of a power line.

Source: Burton (2001)

such rule is that grid connection capacity increases with the square of the voltage level (when voltage level is doubled, wind power capacity can be increased four-fold). According to an official document in Sweden, 3.5MW can be connected to a 10kV line, and 15MW to a 20kV line (and according to this rule 60MW to a 40kV line). In an engineering textbook on wind power published in the UK other figures are recommended (see Table 16.1).

According to the technical rules (so-called grid codes) used in Sweden, the short-circuit power at the point of connection should be 20 times larger than the nominal power of the turbine. There are no harmonized rules on this at an international level, however; each country has its own grid codes. (The grid codes for different countries in Europe are listed in 'Grid connection requirements for wind power technology', a chapter in EWEA, 2005b.)

Local production for local consumption

In the early days of wind power, when the turbines were quite small (20–250kW) they were intended to produce local power that could be used in the near vicinity, and even today many turbines still function in that way. This is an obvious advantage for the power system, since it will reduce the losses in the grid (the further power is transmitted, the larger the losses). As long as the maximum power from the wind turbines is less than the minimum load in the local grid, all their power will be consumed in the local area.

Factories with an even and large power demand can use all the power from a wind turbine. On Gotland, for example, there is a fodder factory in the harbour of Klintehamn that runs day and night all year around, has its own wind turbine (500kW) and feeds the power directly into the factory, where all the power is consumed. (The factory is of course also connected to the grid, otherwise the machines would stop when the wind stops blowing.) A sawmill nearby also has its own wind turbine onsite.

High wind power penetration

When wind power penetration (wind turbines' share of power production) first reaches a level of 10 per cent, the grid and power system may need to be adapted. So far only Denmark has reached this level nationally, but Denmark is part of the larger Nordic power system. In some regions wind power penetrations are already higher: Schleswig-Holstein and Mecklenburg-Western Pomerania in Germany have reached 30 per cent, for example, but these are also integral parts of a much larger power system.

But it is no problem to use a much larger share of wind power in the power system than for the local minimum load. Power systems have good power regulation capacities to keep supply and demand in balance; this is done every minute of the year, with or without wind power, and there are always power plants with reserve capacity in the system (see Box 16.2).

There is always a certain amount of back-up capacity in a power system, since all power plants have limited availability and unplanned shutdowns will occur. If wind turbines are installed in different geographical parts of the grid, it rarely happens that all turbines stop at the same time, since regional variations in the wind are considerable. The wind will almost never disappear at all places at the same time, and much of the variation will be smoothed out. It will not be necessary to build back-up power plants even with quite a large share of wind power in the power system (see Figures 16.5 and 16.6).

The more power systems in Europe are interconnected, the less back-up capacity will be needed. Denmark plans to produce as much as 50 per cent of

Box 16.2 POWER RESERVES

To be able to follow changes in the load (consumption), power reserves have to be available. *Primary reserves* are power plants with very short start-up times, from one second to one minute; these are used for fast regulation. The power plants used for this are gas turbines and hydropower. Power from these is replaced by *secondary reserves*, power plants with longer regulation times, 10 minutes to an hour, and lower operational costs. When the secondary reserves increase their production the primary reserves are replaced.

To prevent a power cut if a failure occurs at a large power plant and it has to close down with short notice, there is always also a *reserve* in the power system. The size of this disturbance reserve is based on the largest power plant that could trip off. In the Nordic power system the disturbance reserve is 1200MW, the size of the largest power plant (a nuclear reactor) in the power system.

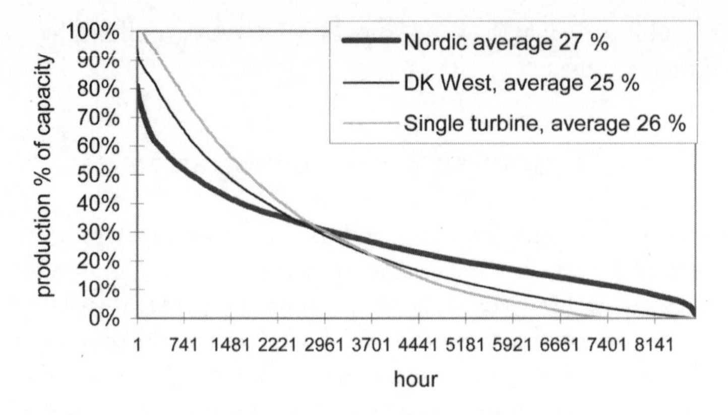

Figure 16.5 *Smoothing effect*

With large-scale development of wind power, when thousands of turbines are installed at hundreds of different sites, the changes in the power productions from wind power will be smoothed out, according to a study made by Hannele Holttinen from VTT in Finland. This diagram, with the hours of the year on the horizontal axis and production in relation to the total installed wind power capacity on the vertical axis, illustrates the smoothing effect. For a single turbine production will vary between 0 and 100 per cent (the steepest curve). For all turbines in western Denmark (Jutland) production will be 0 for just a few hours, and total production will not be over 90 per cent (middle curve). In the whole Nordic system, production below 5 per cent or above 70 per cent of installed capacity will be rare.

Source: Holttinen (2004)

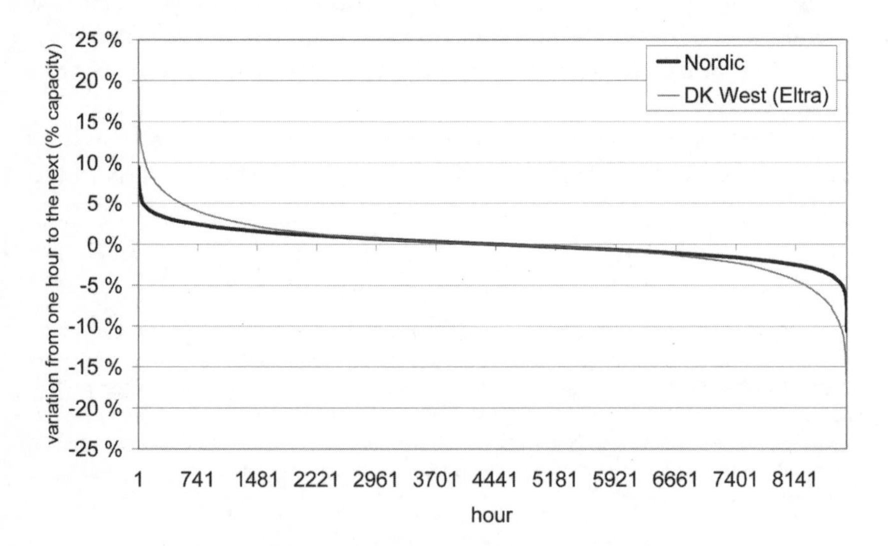

Figure 16.6 *Production changes per hour*

The changes in production within one hour will be small. For 98 per cent of the time the changes within one hour will be within ± 5 per cent of capacity in the Nordic power system (bold curve).

Source: Holttinen (2004)

Table 16.2 *Reserve requirement in the Nordic system*

Wind power penetration	MW	Share of peak load or capacity
5%	80–110	0.8–1.2%
10%	310–420	1.6–2.2%
20%	1200–1400	3.1–4.2%

When wind turbines, geographically well distributed in the Nordic countries, produce 5 per cent of the total electric power, 80–110MW of new power plants for reserve capacity have to be available. With a 20 per cent penetration, up to 1400MW extra reserves will be needed in the Nordic power system.

Source: Holttinen (2004)

its electric power using wind turbines within the next few decades and will manage the balance with the help of hydropower from Norway. This is probably an exceptionally high share of wind power. In most other countries about 20 per cent seems a more practical target. With a high wind power penetration, reserve requirements and costs for wind power integration will increase, but not very much (see Table 16.2).

The value of wind-generated electric power can be increased with good wind prognoses that make it possible to predict power production during the coming hours or days with high accuracy. Prognoses will, of course, not have any impact on the production of the turbines, but they will have a considerable economic value. If predictions and actual production coincide, the penalties for grid operators for deviations from 'ordered' power will decrease (see Box 16.3).

BOX 16.3 WIND PROGNOSES

Computer programs for wind predictions for 6 to 72 hours ahead started to be used in 1997, when researchers at Risoe in Denmark introduced the *Prediktor* software (see www.prediktor.dk; also Risoe National Laboratory, no date). Today there are several different programs in use: Risoe has developed a new one called *Zephyr*, the American company True Wind Solutions (now part of AWS Truewind) offers an internet-based program, *eWind*, and German researchers have developed the program *Previento*.

These programs use input data from national weather institutes or local meteorological stations. These data are processed by statistical methods and are recalculated to derive predictions of the power production of specific wind farms.

Short prognoses for 1–3 hours have an accuracy of 80–90 per cent, but for longer prediction times the accuracy is considerably lower. The prognoses are updated several times a day. Some of these programs are available on the market and are used by wind farm managers or grid operators, while other companies sell prognoses as services for a subscription fee.

Source: Dragza (2001)

When wind power penetration increases above 10 per cent, it will be necessary to have quite good predictions of the power turbines will supply during coming days. The TSO will then also, at least for large wind farms, put the same demands on wind power plants that other power plants have to comply with. This means that wind farms will have to participate in power regulation: they will have to be operated so that they can reduce the power that is fed into the grid or increase power output at short notice. This kind of demand will probably apply only to large offshore wind farms and is possible from a technical point of view, although it will increase the cost of wind power. When cost-efficient means to store electric power, or convert it to hydrogen, have been developed, the integration of very large shares of wind power will be possible.

Remaining barriers

The laws that regulate electric power management differ from country to country. In the EU, however, there are some directives and recommendations that should apply to all member states. Before these enter into force, however, they have to be passed by national parliaments. The EU has taken the decision to deregulate the electric power system and to enable free power trading across national borders. Some countries have already deregulated their power markets, while others have hardly started this process.

On a deregulated market, power production, power trading and power distribution should be completely separated, in other words performed by different companies independent of each other. In many countries, however, huge and vertically integrated power companies (that produce, distribute and sell power), often owned by the state and with a monopoly on the market, still remain.

On the deregulated market all independent power producers have the legal right to get their power plants connected to the power grid if it is technically possible. In several countries the power companies that operate the grid are still very reluctant to let independent power producers connect wind turbines, however. This is sometimes prevented by the introduction of very strict technical requirements on wind turbines. The European Wind Energy Association has described the situation thus:

> *Grid codes and other technical requirements should reflect the true technical needs and be developed in cooperation between independent and unbiased TSOs, the wind energy sector and independent regulators... Grid codes often contain very costly, challenging and continuously changing requirements and are developed in a highly non-transparent manner by vertically-integrated power companies, who are in direct competition with wind farm operators.* (EWEA, 2005)

The capacity of the grid can be a limiting factor for the development of wind turbines. For large projects it will sometimes be necessary to reinforce the grid, which takes a quite large investment. The question of whether these investments should be made by the project developer or by the grid operator is a matter of constant controversy, at least in countries were there aren't strict legal provisions in this regard.

PART V

Wind Power Project Development

To develop a wind power project requires planning, the acquisition of consents, installation and finally operation of the wind turbines. This process includes many different steps that can vary depending on the preconditions. During the feasibility study, described in Chapter 17, the developer will have to decide after each step if it is worth continuing, or if it is better to end the project at an early stage. The most important basis for this decision is the result of the economic calculation, described in Chapter 18. If the preconditions are good enough, the wind turbine has to be sited, or the wind farm designed, to optimize the output, at the same time as the demands from authorities on environment impact and so forth have to be fulfilled for the necessary permission to be given. This process is described in Chapter 19.

17
Siting of Wind Turbines

When looking for a good site for wind turbines many different factors have to be considered. The most important is of course the wind resource. Local conditions like hills and mountains, buildings and vegetation influence the wind and have to be considered in a more detailed calculation of how much wind turbines will be able to produce at the site.

The wind turbine has to be transported to the site, installed and connected to the grid. The distance to existing roads and/or harbours, the costs of building access roads, ground conditions that influence the design and cost of the foundations, and the distance to the grid are thus important factors that have to be included in the calculations.

When the wind turbine has been installed it should not disturb people who live close by. In Sweden and most other European countries there are rules about the maximum noise level that is acceptable and this defines the minimum distance to buildings close to the site. Rotating shadows from the rotor must also be considered and care taken to ensure that they will not be too annoying.

Usually it is also necessary to get permission from the authorities to build a wind turbine. These rules and regulations are specific for every country, but in general the authorities will check that a wind turbine will not interfere or create conflicts with other kinds of enterprises or interests. So it is both wise and necessary for a developer to check what kind of opposing interests there may be at a potential site: airports, air traffic in general (turbines are quite high), military installations (radar, radio links, etc.) nature protection areas, archaeological sites and so forth. Information about opposing interests can usually be supplied by the county or the municipality.

Finding sites with good wind resources

If the task is to develop one or several wind turbines or wind farms within a specified geographical area – a country, region or municipality – the first step is to make a survey of the area to find suitable places, followed by an evaluation to choose the most promising sites for feasibility studies.

The most important precondition for a good wind power project is that there are good wind conditions at the site. Always start by studying wind resource maps for the area, if available. If there are no such maps available, gather information about wind conditions however possible, for example by analysing data from meteorological stations (see Figure 5.5, page 56).

Feasibility study

When a couple of areas with good wind resources have been identified, other preconditions for wind power have to be studied. The following in particular have to be clarified:

- **Neighbours**. Noise and flickering shadows should not disturb neighbours. Can the turbines be sited so that such disturbances can be avoided?
- **Grid connection**. Is there a power grid with sufficient capacity to connect the wind turbines to within a reasonable distance?
- **Land**. Who owns the land in the area? Are there landowners willing to sell or lease land for wind turbines?
- **Permission**. Is the chance of obtaining necessary permissions reasonably good?
- **Opposing interest**. Are there any military installations, airports, nature conservation areas or other factors that could stop the project?
- **Local acceptance**. What opinion do local inhabitants have about wind power in their neighbourhood?

Impact on neighbours

To avoid neighbours being disturbed, a minimum distance of 500 metres to the closest dwellings should be ensured. For a large wind farm this distance may have to be increased. There are rules and recommendations about sound levels, and sometimes also for shadow flicker, that are acceptable at nearby dwellings and holiday cottages. To be able to fulfil these demands, the site where the turbines are to be installed has to be quite large and open. A good rule of thumb is to have a minimum distance of 500 metres for single turbines and a few hundred metres more for wind farms. With such distances the impacts from noise (see Table 13.4, page 159) and shadows will be well within limits (see Table 13.8, page 163).

Grid connection

Power lines are usually indicated on maps, so it is quite easy to estimate the distance from the turbines to the grid. However, it is also necessary to know the voltage level, since that sets a limit to the amount of wind power that can be connected. With a little experience this can be estimated by taking a look at the pylons and the power lines: the bigger they are the higher the voltage. A look at the insulators (which look like a pile of saucers) can give a hint as well; one insulator is 10kV, three 30kV and so on (this applies to older power lines in Sweden). The distance to the closest transformer station is also a decisive factor. To get this information right it is usually best to consult the local grid operator (see Table 16.1, page 212).

Land for turbines

What kind of landowners there are in an area is usually quite easy to guess. In an agricultural district local farmers usually own the land. In that case it's quite likely that it will be possible to find a landowner who is prepared to lease some land for wind turbines. After all, the land can be tilled like before, but there will be extra income for the farmer. Making money out of air is good business. In other cases other private landowners, companies, local communities or the state can own the land. Information on land ownership can be found in the land registry. During the feasibility study it's not necessary to make an agreement with landowners; this can wait until a decision is taken to try and realize the project.

Permission

It's no use spending time and money on projects that can't be built. To evaluate the prospects for getting the necessary permissions from the authorities is one of the most important parts of the feasibility study. The developer has to be familiar with all the rules and regulations that can be applied to a wind power project and how the authorities interpret them. If there are any municipal or regional plans, these can give a good idea of the prospects for the project in an area.

Opposing interests

The possibility of realizing a project can be stopped by so-called *opposing interests*. The first thing to check is if there are any military installations that might be disturbed by having wind turbines close by. Military installations for radar or signal surveillance, radio communication links and so forth are secret, so you can't find them on the map. The developer should make contact with the appropriate military command to find out if they will oppose wind turbines at the site. If this is the case, you can ask them to suggest a place that will not interfere with their interests.

Wind turbines are high structures and can pose a risk to air traffic, especially if there is an airport close by. There are strict rules on how high structures close to the flight paths to and from an airport may be. These rules are available from national aviation authorities.

In most countries there are areas that are classified as areas of international or national interest, to protect nature or cultural heritage, for example: national parks, nature reservations, bird protection areas. Avoid such areas, since it likely will be difficult to get the permissions necessary for wind turbine installations. Such areas are usually indicated on public maps.

Local acceptance

The attitude of the local inhabitants to a proposed wind power project in their vicinity is largely dependent on how the developer performs. In Europe, according to opinion polls and experience, most people have a very positive opinion about wind power (see Chapter 15). At the local level however, there always seem to be some people who strongly oppose wind turbines in their neighbourhood.

How local inhabitants react depends on how they learn about the project. If they get good information at an early stage most of them will be positive. When the developer has decided to realize the project, it's important to create a dialogue with both local authorities and the public and to take the opinions of local inhabitants about distance to dwellings and other practical details into serious consideration. When the turbines are online it is valuable to have local support as local people will then be more inclined keep an eye on the turbines and report any problems.

There are also, however, those who are dedicated opponents to wind power, and organizations for these wind power opponents. Their view is that wind turbines will turn the beautiful landscapes in the countryside into industrial areas and spoil the unbroken view of the horizon at the coast. Even if these opponents are few, they can delay, increase the costs of or even stop projects that are planned by appealing against the building and environment permissions given by the authorities.

This makes it even more important to give proper and good information to all those who may be affected by wind power projects: locals have to feel that they are not being ridden roughshod by the project developers.

Estimating power production

If the feasibility study shows that there is a good site for wind power, one that does not create problems for neighbours, that it is possible to connect to the grid at a reasonable distance, that there are landowners willing to lease or sell land to install the turbines on, and that the prospects to get the necessary permission are good, it is time to make a calculation of how much the wind turbines at the site will produce. The result of this calculation will be the most important input to the economic calculations on which the final decision to go through with the project or not should be based.

Production analysis

A simple estimate can be made directly from a wind resource map (see Figure 5.6, page 57), which shows the energy content of the wind in kWh/m^2/year and also represents specific height agl. Such maps should not, however, be used to calculate a wind turbine's production at a specific site, since they have been calculated with a low resolution. What they do provide is good information about areas where the best wind resources can be found. This distinction between site and area is an impor-

BOX 17.1 ESTIMATION FROM A WIND RESOURCE MAP

If we are interested in a site on a wind resource map for 50m height agl, where the wind energy content is estimated at 4200kWh/m^2/year, and we want to know how much energy a wind turbine with 50m hub height and 50m rotor diameter will produce, we can make an estimation.

With a 50m rotor diameter the swept area will be calculated as follows:

$$3.14 \cdot 25^2 = 1963\text{m}^2.$$

The turbine can be estimated to utilize about 25 per cent ($C_e = 0.25$) of the available energy in the wind. The turbine would thus produce:

$$4200 \cdot 1963 \cdot 0.25 = 2{,}060,000\text{kWh/year}.$$

The estimation of the capacity factor can be based on data from existing wind turbines in the same area, or in areas with the same energy content and hub height. The capacity factor will increase with height.

tant one. Maps can still be used to estimate the production on a site in an area, but this estimation always has to be complemented by a more careful calculation during the feasibility study (see Box 17.1).

To make an exact calculation of how much a wind turbine will produce at a given site, two things have to be known:

1 the power curve of the wind turbine; and
2 the frequency distribution of the wind speed at the hub height at the site.

The power curve shows how much power the turbine will produce at different wind speeds. It is shown as a table, graph or bar chart and is available from the manufacturers (see Figure 17.1). These power curves are verified by independent and authorized control agencies.

It is also necessary to have quite detailed information about the winds at the site. It is not sufficient to know the annual average wind speed but is also necessary to know the wind's frequency distribution; in other words how many hours a year the wind will blow at each different speed (see Figure 17.2).

These data should represent the wind speed distribution during a *normal* year, so average values for a 5–10 year period also have to be recalculated for the hub height of the turbine. Then the power produced at each wind speed is multiplied by the number of hours this wind speed occurs. The frequency distribution is sometimes given as a percentage. If this is the case the percentage value must be simply multiplied by the number of hours in a year (8,760) and divided by 100, to convert it to hours per year.

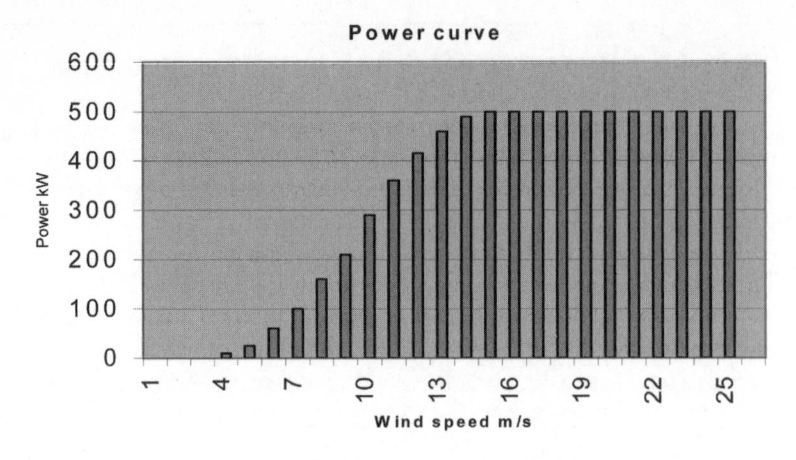

Figure 17.1 *Power curve of the wind turbine (500kW nominal power)*

Source: Tore Wizelius

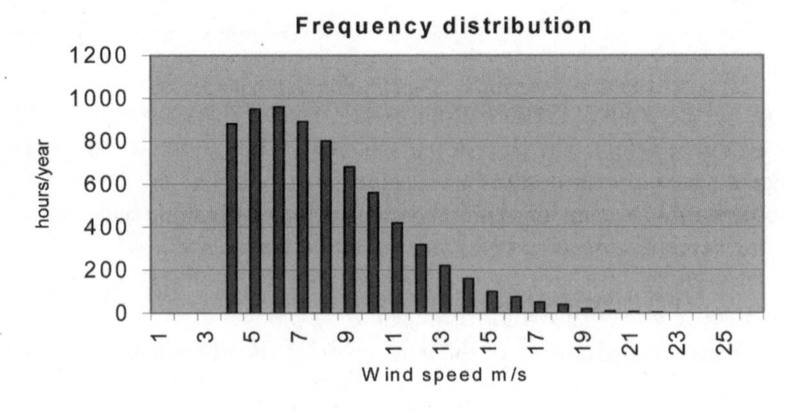

Figure 17.2 *Frequency distribution of the wind at hub height*

Source: Tore Wizelius

How many kWh of electric power the turbine will produce at this site during a normal year can be calculated as the sum of values for the power produced by the turbine for each wind speed multiplied by the hours at that speed. Thus, from values shown in Figures 17.1 and 17.2, annual production = (880 · 10) + (950 · 25) + … + (3 · 500) = 1,227,450kWh.

The wind atlas method

A more accurate way to calculate expected production at a site is to use the wind atlas method, which makes it possible to transform wind data from existing meteorological masts to describe the wind's properties at specific sites within a radius of up to 100km. These data can then be used to make accurate calculations on the expected production at these sites.

Meteorological measurement masts collect wind data in many different places. These data about wind speed and wind direction over time at a given height (10m agl if no height is specified) are representative only for the place where the mast stands, however. Since topography influences wind at each site, these data cannot represent other sites in the vicinity, with other kinds of terrain. To make it possible to use data from one site to calculate the energy in the wind at another site in the same region, the wind data have to be transformed into so-called wind atlas data.

To do this the area surrounding the measurement mast is analysed. The roughness of the terrain is classified, buildings and other obstacles are measured and contour lines are registered. The impact of these factors on the wind is known by practical experiments and measurements that have been generalized into algorithms that are used for these calculations. Wind data from the measurement mast are recalculated by this method to represent wind data for the site that would have been registered if a plain horizontal area with roughness class 1, without any hills or obstacles, surrounded the mast. Finally these data are recalculated for a number of different heights agl (see Figure 17.3).

A wind atlas data set consists of wind speeds (frequency distributions) for 12 different directions (sectors) with one set for each of the following heights: 10, 25, 50, 100 and 250m. In Sweden the state-owned meteorological institute has prepared wind atlas data for some hundred measurement masts from different parts of the country. Wind atlases with wind atlas data are available for most countries in Europe and for many countries on other continents as well. They are available at www.windatlas.dk.

Roughness of terrain

How much a wind turbine can produce depends not only on the character of the terrain at the site, but also on the terrain in an area with 20km radius around it. The terrain conditions close to the site have the greatest influence on the turbine's production, and roughness usually varies in different sectors and thus also with the wind direction.

When a production calculation is made with the wind atlas method, wind atlas data from one or several measurement masts within reasonable distance are used as input. An area with a 20km radius around the turbine site is divided into 12 sectors, and a roughness classification is then carried out sector by sector. Information on obstacles (within 500m of the site), hills and, if the terrain is complex, contour lines are entered into the program (see Figure 17.4).

The wind atlas program then recalculates wind atlas data to wind data at hub height for each sector (see Figure 17.5).

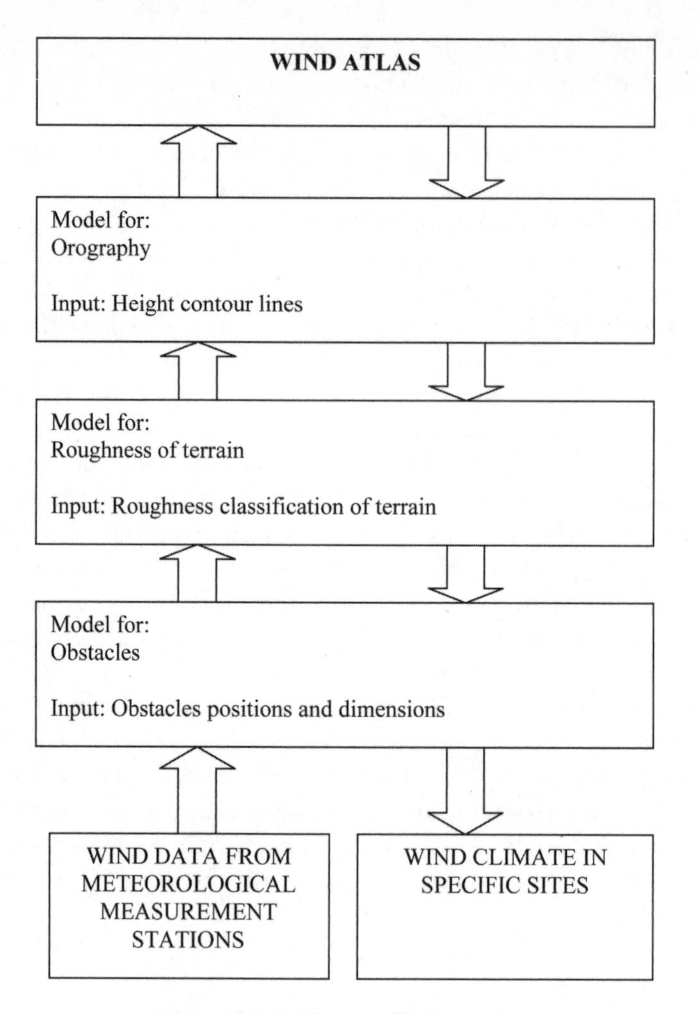

Figure 17.3 *The wind atlas method*

With the help of meteorological models wind data from measurement masts are transformed to data that describe the regional wind climate, wind atlas data. These data about the regional wind climate can then be used to calculate the actual wind climate at a specific site within the region using the same meteorological models.

Source: Troen (1989)

Hills and obstacles

If a wind turbine is sited on the top of a hill or on a slope, this could increase its production. If there are large obstacles close to the turbine, production could be reduced. For large turbines the impact from obstacles is comparatively small, since the impact depends on the difference between the turbine's hub height and the height of the obstacle. The turbulence from an obstacle will spread to twice the

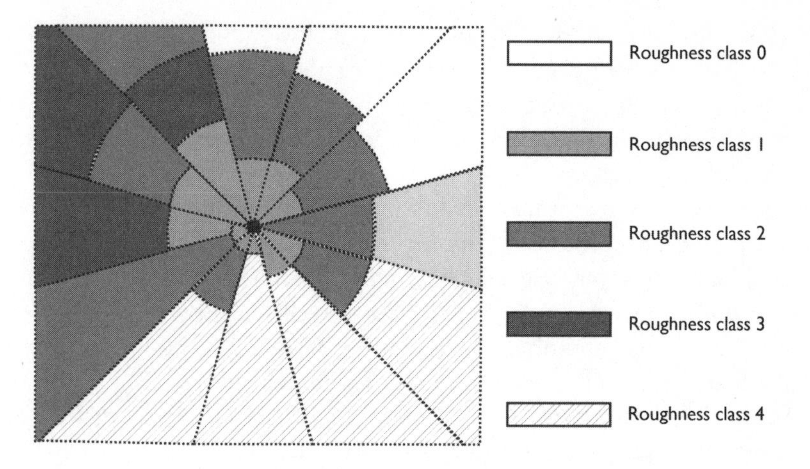

Figure 17.4 *Roughness classification in sectors*

Source: EMD (2005)

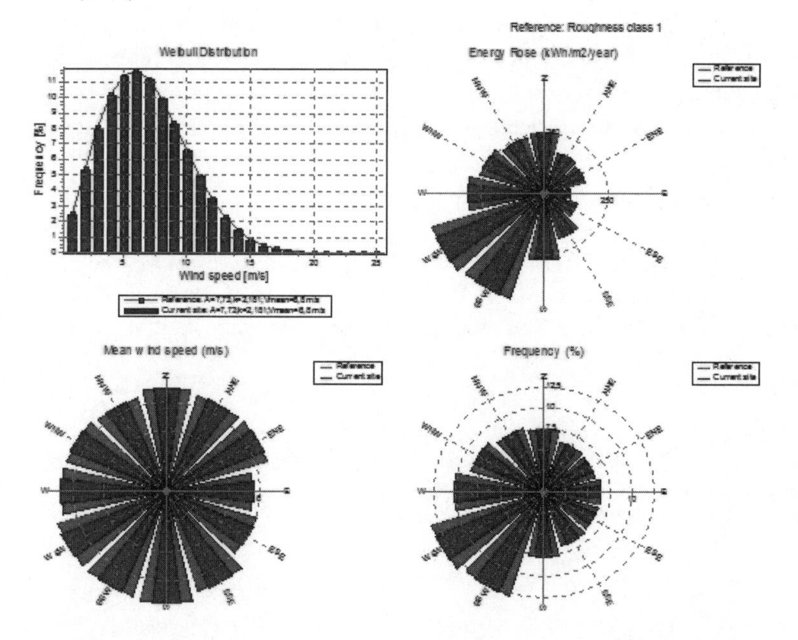

Figure 17.5 *Wind data from a wind atlas program (WindPRO2)*

The program analyses wind data from the site and calculates how much a wind turbine of a specified model, hub height and rotor diameter can be expected to produce there. The frequency distribution at hub height is indicated by a Weibull distribution (top left) and is compared with the distribution for roughness class 1 at the same site. The wind's distribution between different wind directions (sectors) is presented as three different wind roses: the frequency for different directions in per cent (bottom left), mean wind speed for each sector (bottom right) and finally the wind's energy content in the different sectors (top right).

Source: EMD (2005)

obstacle's height. The rotor of a turbine with a 60m hub height and a 60m rotor diameter has its lowest point 30m agl, which means that an obstacle has to be more than 15m high to cause turbulence within the rotor swept area. Low obstacles should be included in the roughness classification and not entered as obstacles in the program in such cases.

The speed-up effect from hills has most impact at lower heights above the hilltop, and this effect increases with the size of the hill. Steep slopes can have the opposite effect, however: if the inclination is greater than 40 degrees, the slope creates turbulence that will decrease production. If the surface is rough or complex, this could happen with an inclination of 10 degrees. A wind atlas program will calculate the impact of hills and obstacles on production.

The wind atlas program first calculates the wind's frequency distribution at hub height for each sector and then multiplies the frequency distributions with the turbine's power curve. The results are weighted according to the frequency for each wind direction and finally summarized. If the terrain is not extremely complex this method gives very accurate results. However, it takes quite a lot of practice and experience of how different kinds of terrain in a region should be classified, along with experience of how far from the measurement mast the wind atlas data are still representative, to make accurate calculations with this method. There are several different versions of software based on the wind atlas method available (see Box 17.2).

BOX 17.2 WIND POWER PC PROGRAMS

There are several different pieces of software for wind power applications.

WAsP

This program has been developed by Risoe National Laboratory in Denmark and is the basis for all wind atlas programs. It can be used to make wind resource maps, wind atlases for whole countries and production calculations for single turbines or large wind farms. Further information is available at www.wasp.dk.

WindPRO

This program can do the same kinds of calculations as WAsP and has additional modules for noise, shadow and visual impact, planning tools and many other functions, as well as a comprehensive database with wind turbine models and wind atlas data for regions and countries. It has been developed by Energi og Miljödata in Aalborg, Denmark. Further information is available at www.emd.dk.

WindFarm

This program has been developed in the UK by the company ReSoft and can do the calculations necessary for project development, including optimization and visualization. Further information is available at www.resoft.co.uk.

WindFarmer

This program can do the calculations necessary for project development, including optimization and visualization. It has been developed by Garrad Hassan and Partners in the UK. Further information is available at www.garradhassan.co.uk.

Greenius

This program has been developed by a German company and can be used for wind analyses and to calculate production and economic feasibility. It can also be used for photovoltaic (PV) and solar heating systems. Further information is available at www.f1.fhtw-berlin.de/studiengang/ut/downloads/greenius/.

Freeware

RETScreen

The CANMET Energy Technology Centre in Canada has developed a comprehensive website with education, databases and simple software for Excel for different renewable energy sources. Further information is available at www.retscreen.net.

ALWIN

This software, developed by the German company Ammonit, which manufactures wind measurement equipment, can be used to analyse wind data and to calculate wind power production. It is available only in German. Further information is available at www.ammonit.de.

Wind resource maps

Wind resource maps can be found at www.windatlas.dk, www.awstruewind.com and www.worldbank.org/astae/, to mention just a few of the interesting sites currently on the internet.

Source: Quaschning (2003)

Sources of error

The accuracy of the calculation depends of course on the quality of the data that are entered into the program. Wind atlas data are based on measured data from different periods (which sequence of years the data are based on is indicated in the database and can vary for different measuring masts). These periods can be too short or not representative enough for the long-term averages. Both wind data and the transformation of these to wind atlas data can be impaired by faults, due to technical faults in the measuring equipment or systematic errors when the data are registered. There is also a certain amount of rounding when data are transformed to the Weibull parameters that are used in the software.

The roughness classification is never absolutely correct, and the roughness can change from season to season and over the lifetime of the turbine. The power curve of the turbine is a third source of error. The form of the power curve depends on the conditions when it was measured and does not give an exact relation between wind speed and power: in different surroundings with different terrain and wind regime it may differ somewhat from the certified one.

These and other factors can be considered to create an error margin of 10 per cent in the calculations, an estimation that has been confirmed by experience. This means that 10 per cent should always be subtracted from the result of the calculation (this is done by software).

Wind atlas programs like WAsP or WindPRO are quite expensive but are a necessity for a professional wind power developer. For single projects, on the other hand, a consultant can do the calculations. Manufacturers also sometimes do the calculations for their customers. When consultants or manufacturers make the calculations, it is important to check that they have taken the error margin of –10 per cent mentioned above into account.

Sound propagation

Wind turbines emit sound. There are rules and regulations about the sound level that is allowed at neighbouring dwellings, which differ from country to country (see Table 13.7, page 161). And the methods (formulas) for calculation differ too. Some countries or municipalities also have rules for minimum distances between wind turbines and dwellings. Some simple programs that calculate sound immission levels at different distances from turbines, based on their sound emission, are available on the internet.

The wind atlas program WindPRO has a module – Decibel – for sound calculations. The sound level at different dwellings can be calculated and maps plotted where sound immission levels at different distances from the turbine are shown. Different country-specific calculation methods can be chosen.

Sometimes it can be hard to find positions for wind turbines that manage to cope with the noise criteria. Several manufacturers therefore offer control systems that can be tuned to different sound emission levels. Operations to decrease sound emission unfortunately also reduce production; such adaptions have a price. On the other hand, sound emission levels can be tailored to the rules, so sound emissions (and production) can be higher during the day than at night, when the restrictions are stricter, for example.

Shadows

The rotating rotor of a wind turbine can create disturbing shadows when the sun is shining and the shadows pass windows in nearby buildings. A shadow diagram will give a picture of how many hours this can occur and during which time of the day. Such a diagram is useful during a feasibility study to avoid disturbing shadows on buildings. The size and form of the areas where shadows can appear depend on the latitude of the site.

For an application and during public consultations more exact information may be needed. The software used for project development often has a tool for shadow calculations as well. The WindPRO module Shadow can, for example, print a so-called shadow calendar, which specifies the exact time and date for shadows on windows for all the houses in the vicinity of a planned wind turbine. The time is calculated by the geometric method (see Chapter 13) and for the worst case, i.e. the sun always shines and the wind always blows from the direction that will give the maximum shadow (rotor parallel to the window). If data on actual sunshine time in different months are available, a more realistic estimation can also be made. The shadow impact can be illustrated graphically as a shadow diagram (see Figure 17.6).

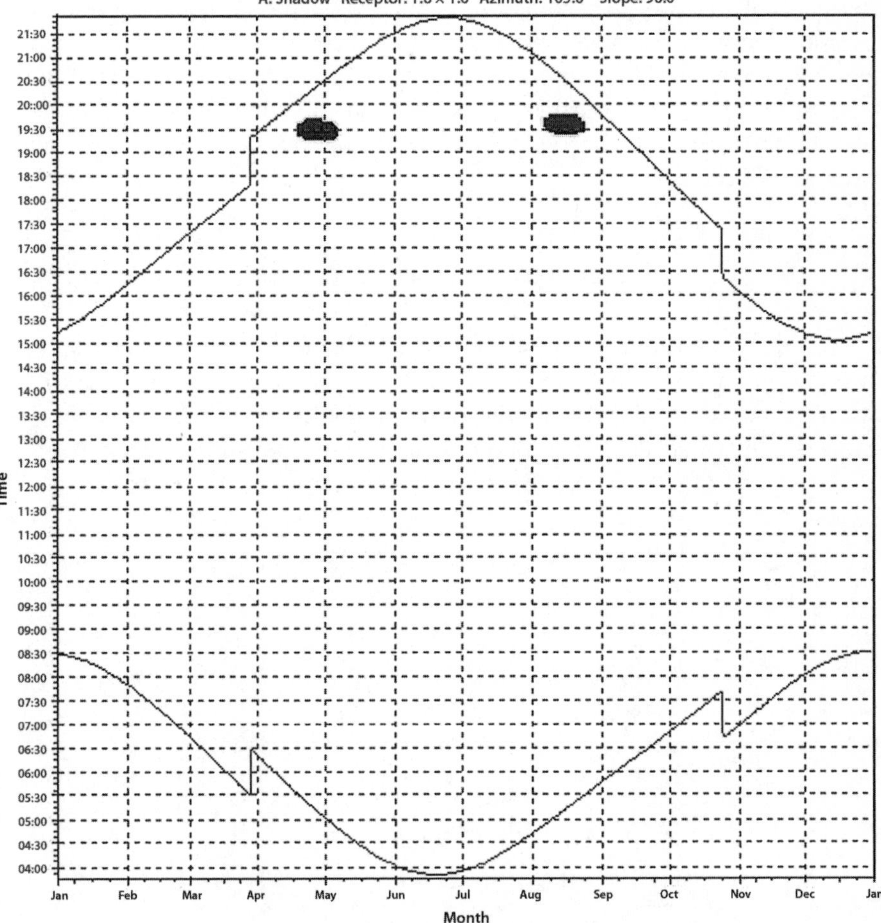

Figure 17.6 *Shadow diagram from WindPRO*

The diagram shows that rotating shadows from the wind turbine rotor will pass the building in the first half of April and in the beginning of September at around 7.30 each evening. The curved lines indicate times for sunrise and sunset.

Source: EMD (2005)

The probable actual shadow time can also be calculated quite accurately. In many countries there are data on the share of sunshine and cloudy weather for different regions. On Gotland, for example, the sunshine time is 25 per cent in a winter month (January) and 60 per cent in a summer month (July). Wind data will give the probability of the wind blowing as well as a frequency distribution for different wind directions. With these data the probability for actual shadows can be calculated. In most cases this probability is 30 per cent (or less) of the 'worst case'. This figure can be used as a rule of thumb.

Such detailed calculations are, however, only necessary if the estimations made show that shadows could become a problem or a matter of dispute. In that case the limitations of the geometric model described in Chapter 13 should be considered as well. If a neighbour risks getting shadow flicker at an unacceptable level, modern control programs can avoid this. A sensor that is connected to the wind turbine's control system registers if the sun is shining and the turbine is stopped during periods (in most cases 10–20 minutes) when annoying shadow flicker would occur.

Groups and wind farms

If more than one turbine is installed at a site, this will have an impact on each turbine's production. How large this impact will be depends on the distance between the turbines and the wind direction. On the leeside of a rotor a *wind wake* is formed: the wind speed slows and only regains its undisturbed speed some ten

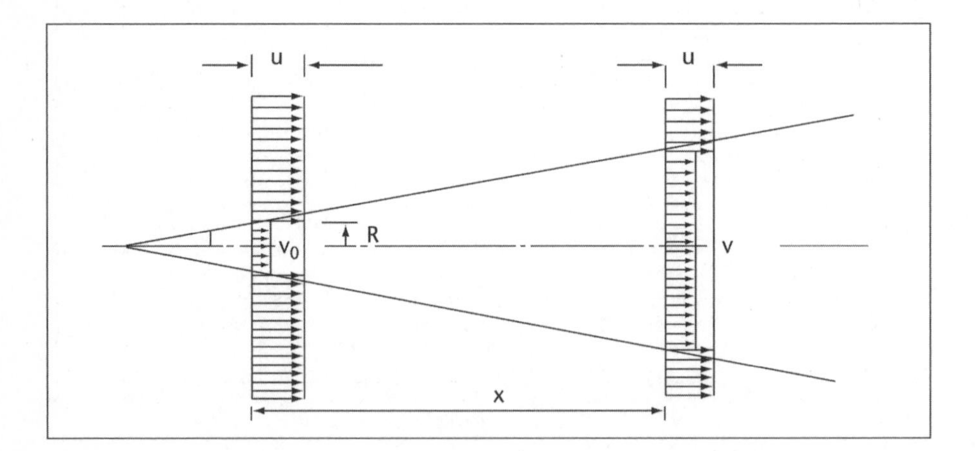

Figure 17.7 *Wind wake*

The wind speed (u) is retarded by the rotor (v_0). Behind the rotor the wind speed increases again (v) as the wake gets wider.

Source: EMD (2005)

rotor diameters behind the turbine (see Figure 17.7). This factor has to be taken into account when the layout for a group of several wind turbines is made.

The wind speed is retarded by the wind turbine rotor and behind the rotor the wind speed increases again until it regains its initial speed. The extension of the wind wake determines how individual turbines will be sited in relation to each other in a group. Behind the rotor the width of the wake increases by about 75 metres per 100 metres and the wind speed will increase with distance. The relation between wind speed v and distance behind the rotor x is described by the formula:

$$v = u\left[1 - \frac{2}{3}\left(\frac{R}{R+\alpha x}\right)^2\right],$$

where:
v is the wind speed x metres behind the rotor;
u is the undisturbed wind speed in front of the rotor;
R is the radius of the rotor; and
α is the wake decay constant (a measure of how fast the wake widens behind the rotor).

The wake decay constant α depends on the roughness class. On land this value is set to 0.075, offshore to 0.04.

In a group of two or three turbines they are usually placed in a straight line perpendicular to the predominant wind direction. The distance between turbines is measured in rotor diameters, since the size of the wind wake depends on the size of the rotor. A common rule of thumb is to place the turbines a distance of five rotor diameters apart if they are set in one row. Larger installations, wind farms, can consist of several rows. In that case the distance between rows is usually seven rotor diameters (see Figure 17.8). In areas where one or two opposing wind directions are very dominant (regions with trade winds for example) the distance between the turbines in rows can be reduced to three rotor diameters.

The actual layout of a wind farm, however, is often dictated by the limits set by local conditions: distance to dwellings, roads and the power grid. If there are height differences on the site, this will also influence how the turbines should be sited in relation to each other to give optimal production. It is usually not reasonable to increase the distance between turbines to eliminate the impact from wind wakes completely as this is an inefficient use of land. The *park efficiency* can be calculated with a wind atlas program. The park efficiency for groups of five to ten turbines usually is 95 per cent or more.

Wind atlas programs calculate park efficiency and there are even tools that can find the most efficient configuration to optimize the production of a specified number of turbines within a limited area. If the area isn't absolutely flat the optimal

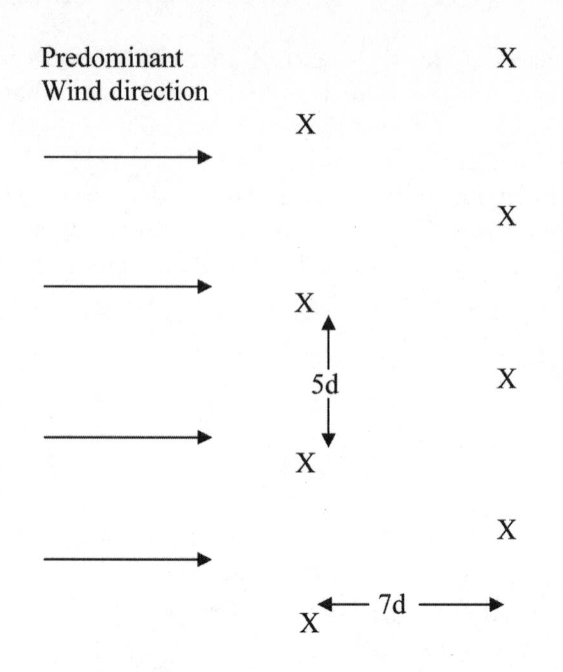

Figure 17.8 *Wind farm configuration*

Source: Tore Wizelius

configuration will be irregular, with distances between turbines differing and the turbines not set along straight lines. In practice, however, the layout is also guided by aesthetic and practical concerns, whether the turbines are set along a coastline, road or headland, in a regular pattern or in a bow like the Middelgrunden offshore wind farm outside Copenhagen (see Figure 3.1, page 26). There they are sited no more than three rotor diameters from each other – much too close to get optimal production – because aesthetic concerns were given a lot of weight.

Visual impact

A good way to illustrate how wind turbines will influence the view of the landscape is to make a photomontage, where the turbines are copied into pictures of the landscape taken from different viewpoints. There are tools for this as well in wind atlas programs, for example the WindPRO module Visual. Pictures of different wind turbines are available in catalogues and can be copied into the pictures at the right scale. Other software, like Photoshop, can also be used to create photomontages (see Figure 17.9).

Figure 17.9 *Photomontage*

This photomontage from Ebeltoft in Denmark shows a row of existing wind turbines on the pier in Ebeltoft harbour and an additional row with five larger turbines that are planned to be built offshore. The viewpoint for the montage is at sea, the background photo was taken from a ferry. The point where the photo is taken is shown on the map below the photomontage, viewpoint A.

Source: EMD (2005)

Wind power in cold climates

There are many places in mountainous areas in arctic regions with very good wind resources. In some of these areas there are already wind turbines installed, in the US and in northern Scandinavia. The arctic climate conditions put special demands on the wind turbines. Ordinary standard turbines would not survive for long, but they can be adapted to the strains that the climate will cause.

Ice build-up

Several different types of ice can be built up on the turbines: rime, fog frost and ice proper. Ice build-up on towers has not caused any problems, but build-up on rotor blades has proved to be difficult matter. Even a very thin ice coating will change the blade profile and thereby also its aerodynamic properties. Efficiency thus decreases and power regulation, especially on stall-controlled turbines, is upset. When this happens the generator will be overloaded and can catch fire.

Ice build-up on the anemometer can also cause problems: this affects the control system, so the turbine may be stopped. For wind turbines to work well in arctic climates, both the rotor blades and the anemometers should be heated. This kind of heating equipment has been developed, and about 1 per cent of the power produced will be used for heating these turbine components.

Extreme temperatures

When the mercury drops down below minus 20°C, the properties of steel and other materials, as well as lubricating and hydraulic oil, will change. Electronic equipment is also sensitive to the cold and if the control system starts to malfunction safety will be at risk. Steel generally becomes brittle in extreme cold, but there are steel types that are able to withstand it.

Cold-adapted grease has to be used in bearings and the hydraulic oil has to be of the same quality that is used in aircraft. The oil and the gearbox have to be heated to working temperature before the turbine can be allowed to start. While they are operating, the low temperature doesn't matter, since the components in the nacelle generate heat as well. To keep the control system computers at a proper working temperature, excess heat from the transformer, if it is installed in the 'cellar' of the turbine, can be utilized.

Extreme heat seems to be less of a problem, since most turbines have cooling systems installed. In tropical climates the humidity in the air can also be very high, however, and may make it necessary to install extra equipment for air-conditioning.

Extreme wind speeds

Most wind turbines are designed to survive extreme wind speeds of at least 60m/s, but only for a few seconds, in other words extreme gusts. For sites where the wind speeds can get higher than that, turbines can be specially designed for such conditions. This will, of course, also increase the cost of the turbines. Several severe storms, hurricanes and typhoons have passed over areas with many wind turbines installed, so the survival ability has been tested in practice. In Denmark most of the thousands of turbines installed survived the severe storms that have occurred during the last years, storms that felled tens of thousands of trees in southern Scandinavia. The turbines that were damaged were old ones, and at fault was neglected maintenance. In India, however, hundreds of turbines were severely damaged by a typhoon in the late 1990s, and in Japan too turbines were felled by a typhoon. In areas where hurricanes and typhoons are parts of the normal climate, wind turbines have to be designed to survive them, even if makes the turbines more expensive.

Offshore wind power

In Denmark, where several thousand turbines have already been installed on land, and it has become difficult to find good sites for more turbines, the first really large offshore wind farms have already been developed, at Horns Rev in the North Sea and at Nysted in the Baltic Sea. Both these offshore wind farms have a capacity of 160MW and both will be further developed in the coming years (this work has already started).

In Germany, which is in the same situation as Denmark, several thousand wind turbines will be installed offshore within this decade. This development is still (in 2006) in the planning stage, but the necessary permissions for several large offshore wind farms have already been approved. In the UK several full-scale offshore wind farms are already online and there are very ambitious plans for offshore development in the coming years.

The wind resources are better at sea than on land. It is, however, also much more expensive to develop wind power offshore. Although several very large offshore wind farms already have been built, the offshore technology has still not reached full maturity. These first large wind farms have had to cope with many unexpected problems with transformers, retrofits of generators and damaged foundations. There are still some teething problems to solve, but that is just a matter of time and experience.

Much effort is being made to find cost-efficient ways to transport and install wind turbines at sea, and manufacturers are competing fiercely to develop ever-larger turbines for offshore installations. A large offshore wind farm can produce as much power as the largest power plants on land. And developing such large wind farms is possible only at sea.

To develop wind power offshore careful investigations of the sea floor have to be made, as well as wind measurements at the sites. The turbines and their foundations have to be dimensioned to withstand the loads from waves and drift ice. To interconnect the turbines and transmit the power to the shore by sea cables also takes advanced and expensive technical solutions. The turbines also have to be designed for minimum maintenance, since it is difficult to get to them far out at sea and waves make it difficult to get 'on board' in bad weather.

Everything, including project management and development, will be more expensive and difficult than on land. However, there is today extensive experience in offshore developments, from the oil and gas fields, so there is no doubt that the problems that arise can be solved when the time is ripe to realize the large-scale development of offshore wind power. How to develop offshore wind farms is, however, outside the scope of this book.

The final step

The purpose of a feasibility study is to find out if the preconditions for wind power in an area or at a specific site are good enough for a successful project. If the wind resources are good enough to guarantee high production, if there are potential sites where turbines will not disturb people who live there or other values, if the prospects of getting the necessary permissions seem good, and if it is possible to connect the turbines to the grid, it is time to continue with the last and most important step in the feasibility study, the economic analysis. The final decision about the project, in other words to stop or to realize it, depends on the result of this analysis. How the economic calculations are done will be described in the next chapter.

18
Economics

To be able to make an informed decision about a wind power project, a thorough economic analysis is necessary. The wind turbines have to generate enough income to guarantee that the investors, or the banks that give loans, will get their money back and a decent return on their investments.

To be able to make a good economic analysis, the first task is to do a realistic calculation of how much electric power the turbines will produce at the chosen site. This is done with a wind atlas program like WindPRO or WAsP or by on-site wind measurements that are normalized to long-term averages. The results of these calculations will tell you how many kWh the turbines can be expected to produce in a normal wind year, i.e. on average during the turbines' lifetime.

The next task is to estimate the *investment costs*, to make a budget for all the investments that will be necessary to realize the project. When the turbines have been installed and start to produce power, the income has to cover the costs and should also generate a profit.

There are several different methods to calculate the returns on an investment: the *annuity* method, the *present value* method and the *pay-back* method. A *cash-flow analysis* illustrates the annual returns on the investment during the turbines' lifetime. All these calculations are, however, fairly uncertain, since they are based on assumptions on future power production (while winds vary), power prices, interest rates and so forth that cannot always be accurately foreseen. Therefore the economic analysis should be followed up with a *sensitivity analysis* that will show the risks and opportunities of the investment.

Finally a plan for the financing of the project that ensures that there always will be enough money in the project to pay interest, repayments of loans and other bills. A wind power project can be financed in different ways – by loans from a bank, by private investors and so forth. It also takes money to develop the project, to build access roads, etc., before any incomes are generated. These costs are usually covered by a building loan from a bank.

Investment

The costs of buying wind turbines, installing them at the site and connecting them to the grid are estimated by an investment calculation, or budget. In a feasibility study, rounded estimations can be used. The purpose of the feasibility study is to find out if the project is worth realizing or not. After the decision is made to actually go ahead with the plans, a new more carefully calculated investment budget is made, based on tenders for turbines and ancillary works (access roads, foundations, grid connection work and equipment, etc.). This carefully calculated investment budget is presented to the bank as part of a loan application and to prospective investors.

The investment budget consists of the following entries:

Wind turbine

Prices of different models and sizes of wind turbines are obtained from price lists or directly from the manufacturers or their agents. During the procurement process prices and conditions can be negotiated. If the turbines are manufactured in another country (with another currency) the price will also depend on the exchange rate, which sometimes can change quite fast. The transport of turbines from the factory to the site, mounting, installation and connection to the grid are performed by personnel from the manufacturer and are usually included in the purchase price. The costs of mobile cranes and some ancillary transport costs have to be covered by the developer. For wind turbines installed on land, the cost of the turbine amounts to about 80 per cent of the total investment cost.

Foundations

The cost of foundations varies a little between different manufacturers. The price for a rock-foundation and a gravity foundation is, however, about the same. The manufacturer will give the technical specifications for the foundations (size, weight, etc.), and the project developer will then ask a local building company to construct them. For offshore installations the foundations are much more expensive.

Roads and miscellaneous

The cost of access roads depends on the size and weight of the turbine, ground conditions and the length of road that has to be built. In many cases it will be sufficient to reinforce existing roads, so trucks and a mobile crane can get to the site. It is often simpler and cheaper to prepare an access road when the soil is hard and dry. When the turbine is online, the road only has to carry an ordinary small van for the service crew. This cost depends on local conditions. The costs for the mobile crane and special transportation costs (by ferry, etc.) have to be covered by the project developer. A mobile crane is usually rented by the day and if the weather is bad (if too much wind makes mounting impossible, for example), this cost can rise fast.

Grid connection

To connect the turbine to the electric power grid, you need a transformer, a cable to the closest grid power line and an electrician to carry out the work. The cost depends on the size and model of the turbine, the distance to the grid and the grid voltage. Large turbines often have an integrated transformer, either in the nacelle or in the base of the tower. The price for the transformer is then included in the price of the turbine. A telephone line to monitor and control the turbine also has to be connected.

Land

If a landowner is to own the turbine installed on his own land, the cost is negligible, just a few square metres will be needed for the foundations and access road. Wind turbines are, however, often installed on land that is leased. In this case a land lease contract is negotiated and signed that will give the turbine owner the right to use the land for a wind turbine for 25–30 years. The landowner will either get a yearly fee or a down payment for the total period when the turbine is installed. The total sum for the turbine's lifetime should be included in the investment budget.

Project development

This includes costs for planning, in other words the time the developer has to spend working on the project, fees for building permission, interest payments during construction and so on. These can all vary quite considerably depending on the time needed for the development process and what fee the developer takes.

It is often hard to calculate the total investment cost correctly before the development actually starts. In a budget you need to get a rough idea and add a margin for unexpected costs. The relative costs for different budget entries, based on actual figures from Denmark, vary considerably for developments on- and offshore (see Table 18.1). These figures and relationships are likely to differ between countries and may change with time.

Figures that can be used for a preliminary investment budget, based on costs in Sweden in 2005 (in Swedish kronor, with rough equivalent values in US dollars and euros), are presented in Tables 18.2 and 18.3. (The exchange rates as of January 2006 were €1 = SEK9.3, US$1 = SEK7.7, SEK100 = €10.75 and SEK100 = US$13.)

Table 18.1 *Cost structure for a wind farm on land vs offshore, %*

	Land	Offshore
Turbine (ex works)	80	40
Foundations	4	23
Electric installation	2	4
Grid connection	9	21
Consultancy	1	10
Land	2	–
Control systems	–	2
Financial costs	1	–
Road	1	–
Total	**100**	**100**

For offshore wind farms the investment costs are considerably higher than for turbines on land. The cost for foundations is much higher, as is that for the undersea cables that connect the turbines to the grid. Figures are based on 600kW turbines in Denmark in 1997 on land and offshore (Tunö Knob).

Source: Redlinger et al (2002)

Table 18.2 *Investment costs for 1MW wind turbine, 60m hub height*

	SEK	€	US$
Wind turbine	8,000,000	860,000	1,040,000
Foundations	600,000	65,000	78,000
Road and misc.	100,000	11,000	13,000
Grid connection*	600,000	65,000	78,000
Land	250,000	27,000	32,000
Project dev.	200,000	22,000	26,000
Total cost	**9,750,000**	**1,050,000**	**1,267,000**

* grid connection fee included.

Table 18.3 *Specific costs per installed kW, 60m hub height*

	SEK	€	US$
Wind turbine	8000	860	1040
Foundations	600	65	78
Road	250/m	27/m	33/m
Grid – cable	350/m	38/m	45/m
Land	250	27	32

These two tables give examples of key figures, excluding VAT. The costs for different entries vary for different manufacturers, nominal powers, rotor diameters, hub heights and local conditions. The table gives an approximate picture of how the total investment cost is divided into different parts. In Sweden in 2001 the average total investment cost for new wind turbines was SEK8126 per installed kW, according to statistics from the Swedish Energy Agency STEM. In 2005 this investment cost had increased to SEK9,750,000 per MW, according to information from Swedish project developers. In 2006 prices increased further due to rising demand and steel prices.

By increasing the hub height the wind turbine will increase its production, since wind speed increases with height. This will, however, also increase the investment cost. To find the optimum hub height from an economic point of view, a simple key figure can be used: total investment cost/calculated annual production (kWh/year). The cost of an increased hub height depends on the size (and therefore weight) of the turbine. The relative cost (cost/metre) will increase with height, since the additional height is added at the bottom of the conical tower, meaning that the diameter of the additional tower sectors increases and consequently also the material needed (see Table 18.4).

Table 18.4 *Relative cost of increased hub height, €/m*

Size	50–60m	60–70m	70–80m	80–90m
600kW	1800	–	–	–
1MW	3000	4000	5250	9000
2MW	–	6000	6500	11,000

Figures are averages from different manufacturers based on a price list from Germany in 2004. The price per metre for increased hub height can vary between manufacturers, but these figures can be used to find an optimum hub height in relation to investment cost and production.

Economic result and depreciation

It is, of course, just as important to find out the economic outcome after the wind turbine has begun to deliver power to the grid. Then the turbine starts to generate income but also draws some costs, and to generate a profit the incomes clearly have to be larger than the costs. To calculate future costs is not so difficult; it is far trickier, however, to calculate the income, which depends on the type of power purchase agreement (PPA).

There are basically two kinds of costs, *capital costs* (interest and repayment of loans), and costs for *operation and maintenance* (O&M). The actual capital costs depend on how the project has been financed. If it has been financed by loans from a bank, the conditions are specified in the loan agreement. If private investors have financed it (for example as a shareholding company or a cooperative), the project will be financed in cash, but the stakeholders expect a good return on their investments.

Commercial wind turbines are designed for a *technical lifetime* of 20–25 years. The actual technical lifetime is not well known, since few turbines have reached that age yet. How much retrofitting is necessary when a turbine gets old is also a factor of uncertainty.

Maintenance costs will, however, increase with age (see Table 18.5). Therefore the *economic lifetime* may be shorter than the technical lifetime. After 15–20 years the maintenance costs may be so high that it makes sense to replace a turbine with

Table 18.5 *Annual O&M cost in relation to investment cost, 600kW turbine*

Year	1–2	3–5	6–10	11–15	16–20
600kW turbine	1.0%	1.9%	2.2%	3.5%	4.5%

The O&M costs for 10–20 years have been estimated from smaller turbines.

Source: Redlinger et al (2002)

a new and more efficient one. In economic calculations the depreciation time is usually set to 20 years, but owners usually opt to pay back loans on shorter terms, in 10 or 12 years. This means that the capital cost (loan plus interest) should always be paid back within 20 years. If the turbine continues to produce power without problems, profits will thus be higher thereafter unless the incomes disappear into repairs and retrofits.

Operation and maintenance

Service

A wind turbine needs regular servicing, just like any other machine. The service crew will make regular checks of the condition, usually twice a year (this depends on the manufacturer). The oil has to be checked and changed every couple of years. The service costs for the first two years are often included in the price, but oil and other materials are not. After that period the manufacturer or a service company offers a service contract, which will cost about SEK40,000 a year (in Sweden) for a 1MW turbine.

Insurance and administration

During the time of the guarantee, usually two years, insurance for fire and public immunity costing about SEK3000/year is needed. When the guarantee runs out, a machine insurance is usually added. The total cost of insurance for a 1MW turbine is then about SEK40,000/year. To own and run a turbine also requires a certain amount of administrative work: invoices have to be paid, as do VAT and other taxes, and the book-keeping has to be taken care of. At least SEK5000/year should be reserved for administrative costs. To have a telephone connected to the turbine costs about SEK2000/year. Then there are usually taxes and fees to be paid. In Sweden there is a fee to the municipality of SEK1000/year and a property tax of SEK32/kW. The grid operator takes a fee for measuring the production that is fed into the grid; in Sweden this costs about SEK7000/year. The O&M cost for a 1MW turbine in Sweden adds up to SEK127,000/year (see Tables 18.6 and 18.7).

Table 18.6 *Key figures for annual O&M costs, 1MW wind turbine*

	SEK	€	US$
Servicing	40,000	4300	5200
Insurance	40,000	4300	5200
Measurement[1]	7000	750	900
Telephone	2000	215	260
Taxes[2]	32,000	3440	4155
Fees[3]	1000	110	130
Administration	5000	540	650
Total costs	**127,000**	**13,655**	**16,495**

[1] by grid operator; [2] property tax; [3] municipality.

Table 18.7 *Key figures for annual O&M costs, specific costs per kW*

	SEK	€	US$
Service	40/kW	4.3/kW	5.2/kW
Insurance	40/kW	4.3/kW	5.2/kW
Measurement		Fixed fee/turbine	
Telephone		Fixed fee/turbine	
Taxes	32/kW	3.4/kW	4.2/kW
Fees		Fixed fee/turbine	
Administration		Independent of turbine size	
Total costs	**15,000 + 12/kW**	**1615+12/kW**	**1940+14.6/kW**

Note that the values in euros and US dollars are values expressed in these currencies of the O&M costs in Sweden and cannot be directly applied for O&M costs in other EU member states or the US since many of the entries are country specific.

Income

The basic income for a wind power installation is the revenue from selling the electric power. The owner has to make a PPA with a power trading company that buys and sells electric power. In many countries the power market has been deregulated during recent years; in others there is still a monopoly. The conditions for a PPA as well as the price per kWh can vary substantially between countries.

There are also special bonuses for wind-generated power in most countries based on the goal of supporting the development of renewable energy sources that have no emissions that harm the environment. Some countries have a *CO_2 tax reduction* (e.g. Denmark), others have *green certificates* (e.g. Sweden, the UK) and others offer a special and *long-term purchase price* for the power (e.g. Germany, Spain). The Kyoto Protocol, an international agreement on measures to control the emissions of greenhouse gases, offers a further opportunity: to get *CO_2 emission credits*.

Although all the countries mentioned above are members of the EU, rules, regulations, conditions, taxes and market situations differ so much that it is necessary to undertake a specific analysis for each. Since rules are changed and future market prices are not known, even this is a very complicated and uncertain task. National wind power associations, state energy agencies or ministries can provide country-specific information on rules, tariffs and so forth.

To calculate the economic result it is necessary to make an assumption about a price per kWh for the coming 20 years. This assumption has to be based on the facts that are known when the calculation is made. Since this calculation will be the basis for the investment decision, it should be supplemented with a calculation for a *worst-case scenario* and also a *best-case scenario*. By doing this you are making a *sensitivity analysis* to estimate the economic *risk*. The higher the estimated risk, the more expensive it is to borrow money for the project: higher risk means a higher interest rate.

In this context it is not possible to make these kinds of detailed analyses for each country. In the examples below SEK0.50/kWh (certificates included) will be used. This corresponds to €0.055 or US$0.065 and is a very low figure – prices in Sweden are low by international standards.

Calculation of economic result

The economic result is the same as the *annual profit* and is calculated thus:

$$P_a = I_a - C_a - OM_a$$

where P_a = annual profit, I_a = annual income, C_a = annual cost of capital and OM_a = annual cost for O&M.

Example 18.1

A 1MW turbine in roughness class 1 will produce about 2,400,000 kWh/year.
The total investment cost is SEK9,750,000.
Price for power (certificates included): SEK0.50/kWh.
Annual O&M cost: SEK127,000.
The annual income is then 2,400,000 × 0.50 = SEK1,200,000.
The annual net income will be $I_a - OM_a$: 1,200,000 – 127,000 = SEK1,073,000.

The investment has been financed by a loan, which gives an annual cost for capital while the loan is being paid back to the bank, including interest.

The annual cost for capital is calculated by the *annuity* method. The annuity is the sum of amortization (pay-back of loan) and interest where the sum of the amortization plus interest will be constant, i.e. the same each year. The annual capital cost C_a is calculated by the so-called annuity formula:

$$C_a = a\, C_i$$

where a = annuity and C_i = investment cost;

$$a = \frac{rq^n}{q^n - 1}$$

where r = interest rate, n = depreciation time in years, and $q = 1 + r$.

If the interest rate is 6% and depreciation time 20 years, then:

$$a = \frac{0.06 \times (1.06)^{20}}{(1.06)^{20} - 1} = 0.087185$$

$$C_a = 0.087185 \times C_i = 0.087185 \times 9\,750\,000 = \text{SEK}850{,}054.$$

The annual profit P_a = net income – annual capital cost C_a:

$$P_a = 1{,}073{,}000 - 850{,}054 = \text{SEK}222{,}946$$

Present value method

Another method to calculate the economic result for an investment in wind turbines is the *present value method*, also called the *discounting method*. With this method the value of an annual income or expense that will occur for a specific number of years is given the value at a specific time, usually the day when the turbine starts to operate. If the present value of the revenues is larger that the present value of the investment and expenses, the investment will be profitable.

The present value is calculated by this formula:

$$PV = f_c \times R$$

where PV = present value, f_c = capitalization factor and R = revenue.

$$f_c = \frac{q^n - 1}{r \times q^n}$$

where r = real interest rate (interest – inflation), n = number of years and $q = 1 + r$.

Example 18.2

Assuming an operation time for the turbine of 20 years and a real interest rate of 6%,

$$\text{capitalization factor } f_c = \frac{1.06^{20} - 1}{0.06 \times 1.06^{20}}; \quad f_c = 11.5.$$

The present value of the net income is:

$$N = 11.5 \times 1{,}073{,}000 = \text{SEK}12{,}339{,}500.$$

The profit during 20 years of operation will thus be the present value of the net income minus the investment cost:

$$P_{20} = 12{,}339{,}500 - 9{,}750{,}000 = \text{SEK}2{,}589{,}500.$$

Pay-back method

A third method to evaluate the economic preconditions for an investment in a wind power project is the *pay-back method*. This is used to calculate how long it will take to get back the money that has been invested. The pay-back time is calculated with this simple formula:

$$T = \frac{\text{Investment}}{\text{Annual net income}}.$$

Example 18.3

The pay-back time for the wind turbine in Example 18.1 will be:

$$T = \frac{9{,}750{,}000}{1{,}073{,}000} \approx 9 \text{ years.}$$

What is the production cost for wind power?

It is also possible to calculate the actual cost to produce 1kWh with wind power. This energy cost is equal to the annual capital cost plus the annual O&M cost divided by the annual production in kWh.

$$E_{\text{cost}} = \frac{C_a + OM_a}{\text{kWh/year}}.$$

Example 18.4

With a technical lifetime of 20 years (power production) and an interest rate of 6% (real interest = interest rate + inflation), the annuity will be a = 0.087185. Annual capital cost will become C_a = SEK850,054.

$$\text{The cost for energy } E_{cost} \text{ is then } \frac{850{,}054 + 127{,}000}{2{,}400{,}000} = \text{SEK}0.41/\text{kWh}.$$

According to all these calculation methods the wind power project used in these examples will be a reasonable investment, at least in Sweden, but probably also in most other countries. The calculations are not based on expectations of rising power prices, but only on actual prices today and real interest rates (interest rate minus inflation). They give a good indication about economic viability, compared to other similar projects, and can also be used to compare the economics of different turbine models (sizes, manufacturers, etc.) that could be used in the project, or to calculate if it makes sense to increase the hub height, a measure that will increase power production but also the investment cost.

Risk assessment

The calculation of the economic result is based on several assumptions. The first is the total power price. The second is the calculated power production. What happens if these assumptions are wrong? To find this out it is always wise to make a risk assessment through a sensitivity analysis. A scenario is made for the worst things, within reason, of course, that can happen. In this case, for example, that the total power price will sink to SEK0.45/kWh and that the power production will be 10 per cent lower than calculated (due to errors in the calculations, climate change or other reasons). These figures are then used in the same calculations as before. If this results in figures going into the red – an annual loss instead of an annual profit – then the project can be classified as high risk.

However, a similar best-case scenario should also be made, for example that the total power price will rise to SEK0.55/kWh and power production will be 10 per cent higher than calculated (due to errors in the calculations, climate change or other reasons). Thus a chance and a risk are equally presented, although evaluating these is a matter for the credit institute and the investors.

None of these calculation methods, however, is sufficient to work out a real and realistic budget for a wind power project. To do that a much more detailed analysis, based on tenders, actual credit conditions from banks and so forth, has to be made. Interest rates, for example, have a large impact on economic viability and are influenced by the perceived risks and opportunities of the investment.

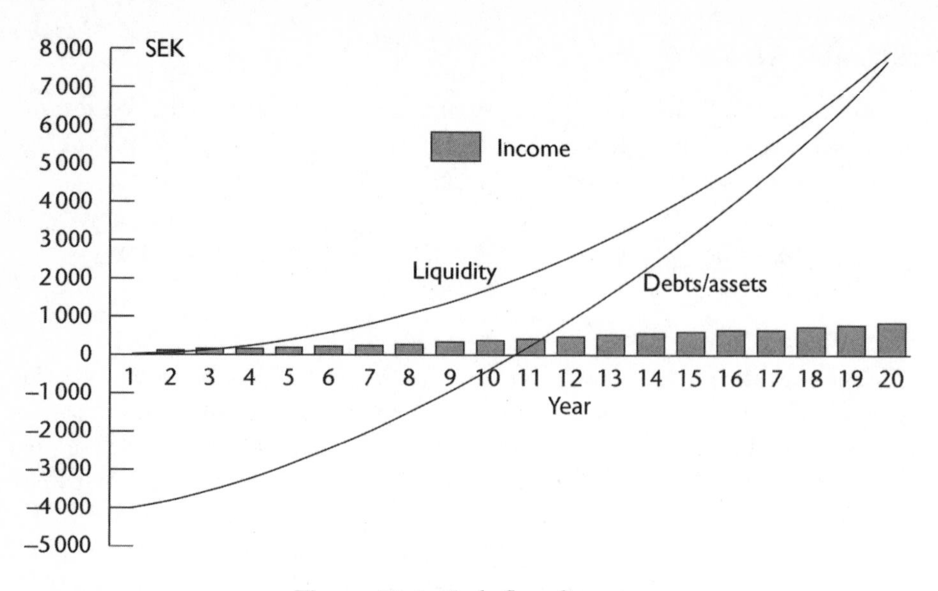

Figure 18.1 *Cash-flow diagram*

A cash-flow diagram shows the variations of the cash flow over time, i.e. the relationship between debts (loans), incomes (assets) and liquidity (cash assets). This diagram shows that this project has a surplus and positive cash assets and that the pay-back time will be 11 years (where the debts/assets graph crosses the x-axis. (Note: this diagram is *not* based on the figures in examples 18.1–4.)

Source: STEM (1999)

Cash-flow analysis

A so-called cash-flow analysis is a good method to calculate the economic result year by year. This shows the cash flow during the economic lifetime of the turbine and can be made with software like Excel. Information on calculated power production, power price, green certificates and other bonuses, loans, interest rates, and other factors that have an impact on the project economics are entered into the spreadsheet. Expected inflation rates and increases of power purchase prices can be entered as well. The program then calculates the outcome year by year and cash flow; annual revenues, capital costs, maintenance costs and remaining surplus can be presented in diagrams and tables (see Figure 18.1).

Sensitivity analysis

The cash-flow method can also be used for a so-called sensitivity analysis, a calculation of how earnings are influenced by changes in the power price, green certificates, interest rates and other factors. The calculation of the economic result is based on assumptions of the price per kWh that is produced, using the price that

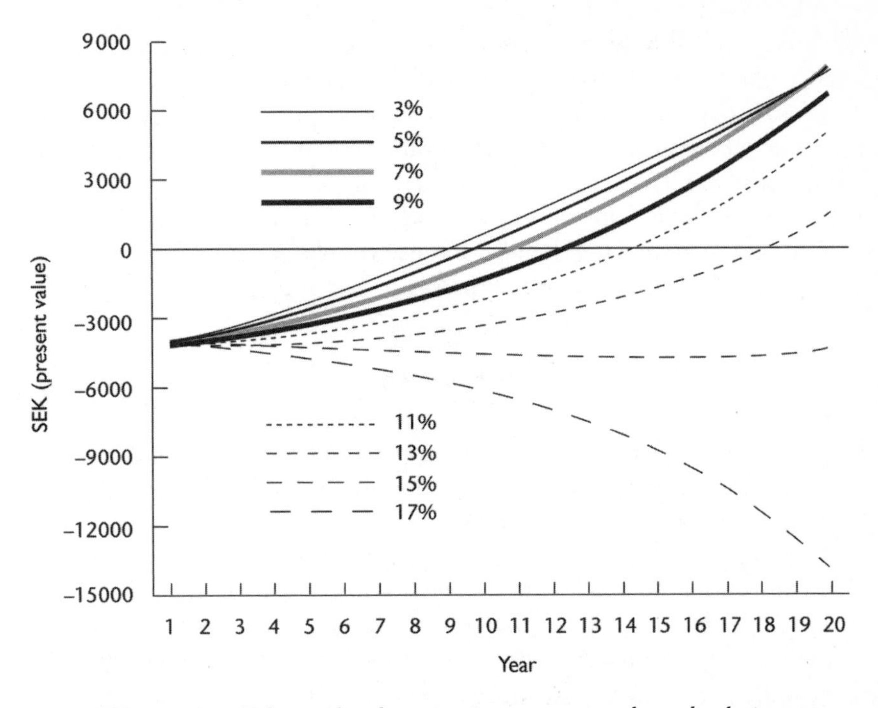

Figure 18.2 *Relationship between interest rate and pay-back time*

The relationship between the interest rate on loans and pay-back time varies for different projects and depends on investment costs, calculated production and other factors. The higher the interest rate is, the longer it will take to pay back the loans.

Source: STEM (1999)

seems most likely. Let us presume that we have judged a price of SEK0.5/kWh as most likely in our normal case. This is the best qualified guess we can make, but what happens if it is too optimistic? The calculation for the normal case should therefore be complemented with a calculation for the worst case, say SEK0.45/ kWh. And how will the economic result turn out if the power price increases to SEK0.55/kWh?

The interest rate has a great impact on the economic result. The higher the interest rate is, the longer it will take to pay back the loans. The relationship between the interest rate and the pay-back time can also be illustrated by a cash-flow diagram (see Figure 18.2).

The uncertainty over the energy content of the wind should be considered. The economic calculation for the normal case is based on the power production during a normal wind year, taken from a 5–10 year average. But for any specific year in the 20–25 years that the wind turbine will operate, the energy content in the wind can be considerably different. In some of these years the turbines could produce 20 per cent less than average, others 20 per cent more. The worst case is two consecutive years of less than normal wind just when the turbines have started their operation. With this calculation it is possible to check that the revenues will be able to cover the operation costs if this happens.

There is one more factor to consider regarding project economics. In many countries there are quite large seasonal variations in the wind. Even if the economic result on an annual basis looks good, there can be deficits during the year, after a month when production has been low and there are bills to be paid. Therefore it is a good idea to work out a liquidity budget as well, to make sure that there will be cash available whenever money is needed.

Financing

How a wind power project is financed depends on what kind of owner it has. There are many different kinds of ownership that can be involved, in Sweden as well as in other countries (see Table 18.8).

Big companies may have the capital needed for the investment available within the company. This is called *corporate financing*. The other financing principle is called *project financing* and can be utilized by large corporations as well. In this case the wind power project is treated as an independent economic entity. Small and medium-sized enterprises, which have been formed for the sole purpose of owning and operating wind turbines, usually have to take out loans from a bank or other credit institution. In a wind power cooperative the members, who have to pay for their shares in the association, will finance the investment in cash. A limited company will raise some or all of the money for the investment from equities.

A *financing plan* consists of a type of flow-chart that shows the expenses of the project, when invoices have to be paid and where this money should come from. To make such a plan it is necessary to have discussions and negotiations with the

Table 18.8 *Wind power divided into owner categories, Sweden*

Category	Share (%)
Private person or firm*	18.1
Wind power cooperative	9.7
Wind power company (Ltd)**	41.3
Enterprise outside energy branch***	6.8
Utility, Power company	19.9
Unknown	4.2

* one person firm;
** wind power as main business;
*** agriculture, shop, etc.

Share of installed nominal power in 2004, divided into different owner categories.

Source: Elforsk (2005)

supplier, entrepreneurs and the bank. If the wind turbines are imported from other countries (with other currencies) it is usually a good idea to ensure (i.e. fix) the exchange rate, since even minor changes in a currency can have large impacts on the economics of a project.

19
Project Development

If the economic analysis shows that the planned project will be a good investment, it is time to take the steps necessary to realize the plans.

The area where the turbines are to be installed has been identified in the feasibility study, now the exact location of the turbines within this area has to be decided. Usually there are several other factors to consider: how much power can be connected to the grid, specifications about minimum annual production, maximum investment costs and demands on economic return from the customer (utility, power company, etc.). The developer's task is to plan an optimized wind power plant within the limits of these conditions and restrictions.

Project development consists of the following steps:

- **Early dialogue**: inform local authorities, neighbours, etc.;
- **Land acquisition**: negotiate a contract with landowners;
- **Detailed planning**: decide number and size of and sites for turbines;
- **Second dialogue**: present the detailed plan to the authorities and public;
- **EIA**: work out an environmental impact assessment for the project if necessary;
- **Permission**: apply for building permission etc.;
- **Purchase**: ask for tenders and choose the best offer;
- **Contracts**: sign agreements with grid operator and power company/utility;
- **Installation**: install turbines and connect them to the grid; and
- **Transfer**: transfer the wind power plant to buyer/owner.

The wind power project should give the best possible return on the investment, but it also has to be compatible with the demands of authorities so that necessary permission will be granted. The project development process, as well as the purchase, has to be financed. This is another task for the project developer to work out (see Figure 19.1).

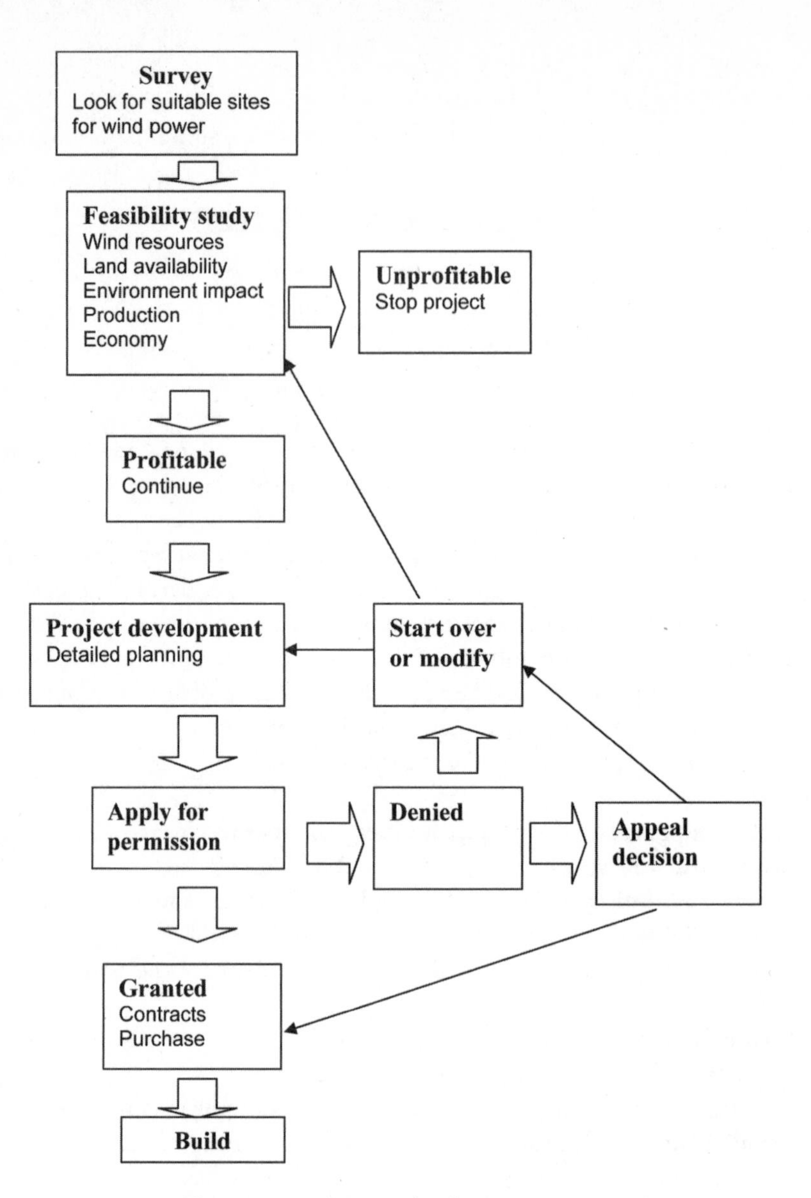

Figure 19.1 *Project development process*

Source: Tore Wizelius

Early dialogue

The developer can start by making rough outlines for a few different options for a wind power installation and invite people in the surrounding area (1–2km from the site) to an information meeting, an *early dialogue*. Local and regional authorities, the grid operator and the local media should also be invited. The developer can explain about wind power in general, the environmental benefits, local wind resources and impact from noise, shadows and so on, and finally show some outlines and ask participants for their opinions. Representatives from the local and/or regional authorities can state their opinions about the proposed project and describe how a decision will be taken.

The developer should also have an early dialogue with the local community, the grid operator and other relevant authorities in separate meetings. The project should be a rough outline: the point of an early dialogue is to adapt and modify the project to avoid unnecessary conflict.

Land acquisition

Access to land is necessary to install and operate wind turbines, so an agreement with landowners should be made at an early stage. If more than one landowner is involved, a common agreement should be made, although the land lease contracts will be individual. In Sweden the lease usually is set at 2–4 per cent of the gross annual income from the turbine, or a corresponding sum paid up front when the turbine is installed. The terms of a land lease contract are a matter for negotiation between the landowner and the developer, but it is wise to make a fair deal that is in accordance with other similar contracts. It is always valuable to have someone living close to the site that has the turbines under surveillance.

Detailed planning

The developer's task is to optimize the wind turbines within the limits set by the local conditions. To find the best solution, wind turbines of different size (height and rotor diameter) and nominal power should be tested theoretically at different sites within the area. This process is called *micrositing*. For these different options the production should be calculated and the economics analysed. The impact on neighbours and environment also has to be checked. Finally the developer has to choose the best option.

In a wind farm with several turbines, the layout (number, size and configuration) should be adapted to the local wind conditions, so the wind resources are used in the most efficient way.

In practice, though, there are always *boundary conditions* to consider. These conditions are defined by dwellings (minimum distances due to noise regulations, etc.), other buildings, groves and other obstacles, roads, the power grid, topography, property borders, coastlines and so forth. Boundary conditions limit the area that is available for wind turbines.

A developer who plans to install a 10MW wind farm can choose ten turbines of 1MW, five turbines of 2MW or some other option. The economic outcome should, of course, be optimized, but the risk of getting an application turned down should at the same time be minimized.

With the aid of know-how, good judgement, a constructive dialogue with neighbours and local authorities, and good wind data and PC software, the developer will find the best solution for the project, a detailed plan that should be realized.

EIA for wind power

For large wind power projects an EIA can be demanded. In the EU there are some common rules on EIAs, but also rules and recommendations that are country specific. The definition of a large project, in other words when a proper EIA has to be made, varies. In Germany this limit is set to 20 turbines or more, while in Sweden an EIA has to be made for any installation of more than 25MW.

An EIA consists of a *process* and a *document*. Dialogues and consultations with the local inhabitants, authorities and other parties that will be affected by the project make up the process. The purpose of this process is to give these groups an opportunity to influence the design of the project so that the impact on the environment will be minimized. The EIA document describes this public consultation process as well as impacts on the environment during construction, operation and dismantling. The EIA process should begin as early as possible so that the parties that will be affected will get a real opportunity to exert an influence on the project design. It is not sufficient to merely write a report when all the details have been firmly fixed.

Information for local inhabitants

In many countries wind power developers have established a practice for planning that is in accordance with the intentions of the EIA process. Most developers organize local information meetings at an early stage to try to ensure that the public will be well informed and have a positive attitude to the plans. Sometimes they are also offered the opportunity to buy shares in the wind turbines.

This information meeting is also the first step in the EIA process (early dialogue). The developer has to present several different options for the siting of turbines and also discuss practical matters pertaining to the construction process – the building of access roads, power lines and so on. A so-called zero-option, in other words the consequences if the project will not be built, also has to be shown. The developer can of course argue for the preferred option, but should be sensitive to other opinions that are put forward. The fact that local inhabitants know the area they live in very well has often proved to be useful to the developer.

By this dialogue the project is made concrete and is designed to minimize impacts on the environment and neighbours. It is then time to start with the EIA document.

The EIA document

An EIA describes the impact of wind turbines on the global, regional and local environment. A detailed description of impacts from sound propagation, shadows and visual impact should always be included.

Each impact (sound, shadows, landscape, etc.) is described on three levels:

1 present situation;
2 impact (change, consequences); and
3 precautions (measures that minimize impacts).

Furthermore, the impact during different stages of development should be described:

- building stage: preparatory work, access roads, power lines, working area for cranes, excavators, trucks, storage, etc.;
- operation stage: visual impact, sound propagation, shadows, safety; and
- restoration: how the turbine will be dismantled and the ground restored.

At least two different options for siting and/or design of the project and a so-called zero-option are included in an EIA.

The different options and their environment impacts are described in a way that makes it possible to compare them and to assess which option will be best for the environment. The zero-option describes the consequences if the project is not realized: it could describe how the electric power that would be produced by the turbines will be otherwise supplied (by coal or natural gas and the consequent emissions), for example. The zero-option does not necessarily imply that the area will be preserved in the same state as at present: it might describe some other kind of change, the land becoming overgrown, for example. The following sections should be included in an EIA.

Introduction

EIA documents are often focused on local impacts. For wind power projects it is, however, important to stress the positive impacts at the global and regional levels, in other words that the emissions of greenhouse gases that can cause climate change, and of sulphur and nitrogen oxides that cause acidification and eutrophication, will be significantly reduced. These reductions should be quantified.

Summary

If the EIA report is very comprehensive, a summary should be included at the beginning of the document.

Project description

A concise description of the project: place, number of turbines, access roads and power lines for grid connection.

Consequences

A description of the consequences of the different options, including the zero-option: the consequences for health and safety (noise, risk of accidents), the environment (landscape, flora and fauna), views, recreation, cultural heritage and natural resources should all be accounted for.

The following elements are described for the different options.

Sound propagation

Distances to sound immission levels 45, 40 and 35dBA (or others that are relevant according to rules and regulations) are given, with these zones indicated on a map. Calculated noise immission levels at dwellings and other building in the vicinity of the turbines are presented. The calculation method recommended by the national authorities has to be used for this (see Chapter 17).

Safety

The risk that ice will be thrown from rotor blades and other accidents is extremely small. At the minimum distance to dwellings defined by the rules for sound immission the risk of accidents is negligible.

Nature

The physical impact on the natural environment is restricted to the building site and can be reduced to a minimum by good planning. When it comes to wildlife, the impact on birds has been much debated, with much research conducted. This has shown that at normal sites wind turbines do not have any significant negative impact on birds.

Visual impact

How the wind turbines will look from different viewpoints in the area can be illustrated by photomontages. Zones of visual impact, areas in the landscape from which the turbines can be seen, can be calculated and shown on maps.

Recreation

Wind turbines are normally not fenced in and do not restrict the access of the public to the areas where they are installed. Neither are there any reports that imply that people would avoid areas with wind turbines for recreation (see Chapter 15).

Cultural heritage

If windmills once utilized the wind resources in an area, this can be used as an argument for developing wind power, since it conforms to a local tradition. In areas with relics of antiquity, there is a risk that new relics will be found when the foundations are built. In that case the developer could be obliged to stop the development and pay for excavation by archaeologists before the project can be built. A distance of respect always has to be kept to churches and other historical buildings.

The marine environment

For offshore projects additional topics have to be investigated in an EIA, covering impacts on:

- fish and other marine organisms;
- migrating and resting birds;
- fisheries; and
- shipping.

Comparative assessment

When the different options and their consequences have been described, a comparative assessment has to be made, with the preferred option backed by convincing arguments. Note that the zero-option should not just describe the area as it stands, but how it may develop during the coming 25 years if wind turbines are not installed. It is also important to point out that the area can be restored to its original state after the turbines have been dismantled.

Precautions

Measures that will be taken to prevent, reduce or compensate for impacts are presented in a separate section. Here, principally, measures to prevent damage during the building process should be described.

Conditions, follow-up and inspection

When it comes to wind power it is chiefly sound immission levels at dwellings that are checked. An inspection programme is often included as a condition for permission being granted.

Public consultation

A short summary of the process of public consultation and dialogue with the authorities on which the EIA report is based should be included.

Sources

Sources for facts and figures, references, and who is responsible for the assessments that are presented should all be included.

Appendices

Expert reports, calculations, maps and so on are provided in appendices.

Building permission

In Sweden an application for building permission is submitted to the local municipality. For projects of more than 1MW an application also has to be submitted to the regional authority (to the environment court for projects of more than 10MW).

In many other countries the only permission needed, up to a certain size of a project, is building permission from the municipality. It is not necessary to present a full EIA for building permission, but is usually sufficient to describe impacts on neighbours (noise and shadows).

Often it is possible to get advance notification from the building committee. It is always advisable to apply for such advance notification, since it will give the developer a good idea about the attitudes and apprehensions among local politicians and make it possible to adapt the project to avoid the final application being rejected.

In an application the turbine manufacturer should not be specified, just the size of the turbines (hub height and rotor diameter) and the sound emission from turbines of that size, otherwise the developer will be bound to a specific manufacturer and this will impair the negotiation position in procurement. If the process to get permission takes a long time, the model applied for may no longer be available on the market, since wind turbine models are continuously upgraded, so rotor diameters and other technical specifications may have to be changed. Thus when permission is finally granted, it may no longer be valid and the whole process may have to start all over again (this has happened several times in Sweden). A sensible solution to this is to add a few metres to hub height and rotor diameter in the application.

Appeals

After the relevant authorities and political bodies have processed the application and the developer has eventually got the necessary permissions, it is another couple of weeks before these become *unappealable*. After that the actual building of the wind turbines can start.

During this period, however, a neighbour, holiday cottage owner or even an authority may raise an appeal against the decision. In this case the developer will have to wait until the court has tried the appeal.

Such legal processes can delay a project for several years and sometimes even put a stop to it. This risk is another good reason to inform all concerned parties and adapt the project to avoid nuisances, even if so doing reduces the economic results a bit. If the permissions are appealed, the costs will in most cases be much larger than such small losses.

Purchase

When all necessary permissions are granted, it is time to purchase the turbines and other goods and services necessary to realize the project. Always ask for tenders from several different suppliers. Use local companies and entrepreneurs to build access roads, foundations and so forth. Evaluate the different tenders and sign a contract for the one that is most favourable. This is not always the one that offers the lowest price: the supplier's record, ability to carry our maintenance and other factors should also be considered.

Contract

Contracts have to be signed with landowners, the grid operator, and a power company or utility that will buy the power. Contracts for credits from banks and other financial institutions also have to be signed.

Installation

The developer has to prepare the site and build the foundations. Gravity foundations of reinforced concrete have to harden for a month. The turbines are usually mounted and installed by the turbine supplier. To mount a large turbine does not take more than one day. Installing the transformer (if it is not integrated in the turbine) and routing a power line to the grid are down to developer.

Transfer

When the wind turbines have been mounted, connected to the grid and thoroughly tested by the supplier and are ready for regular production, it is time to hand them over to the client, unless the developer intends to own and operate them himself.

Operation

Once the wind turbines have been installed and connected to the power grid and have started to operate, they will operate unattended. The owner or the person in charge of the operation can keep them under surveillance from an office, since the turbine's control system is connected by modem to the operator's PC. Simple operational disturbances can be attended to from a distance, and the turbine can be restarted from the PC. When more serious disturbances occur, the operator has to go to the turbine to attend to the fault before it can be restarted. When a fault occurs, the operator will get an alarm on his cell phone, staff locator or PC. Regular servicing is usually carried out twice a year, although this may vary for different models.

How much power the turbines produce is registered on a meter from which the grid operator takes readings. This information is also conveyed via modem, and production is read once a day. Settlements are usually made once a month, when the owner gets paid for the power that was delivered to the grid the preceding month.

Dismantling

When the wind turbine, after some 20–25 years of operation, is worn out, it will be dismantled. Most of the parts can be recycled as scrap metal. The only components that can't be recycled today are the rotor blades, but there are efforts to find methods for that too. The scrap value of a turbine is about the same as the cost of dismantling it. The foundations of reinforced concrete, built below ground level, can be left behind, if they do not affect ground conditions in a negative way. Otherwise they can be removed and reused as hardcore for roads or buildings. Once the turbines have been dismantled, no trace of their presence remains. On a good site, however, a new generation of wind turbines will be installed. This regeneration process has already started in Denmark and Germany.

Acronyms and abbreviations

agl	above ground level
BRP	balance responsible player
CHP	combined heat and power
CO_2	carbon dioxide
dBA	decibel A (weighted)
EIA	environment impact assessment
EWEA	European Wind Energy Association
GIS	geographical information system
GW	gigawatt
G2	second generation
HAWT	horizontal axis wind turbine
IPP	independent power producer
kW	kilowatt
MCP	municipal comprehensive plan
MW	megawatt
NFFO	Non-Fossil Fuel Obligation (UK)
NIMBY	not in my back yard
NOx	nitrogen oxides
O&M	operation and maintenance
PPA	power purchase agreement
R&D	research and development
SMHI	Swedish Meteorological and Hydrological Institute
SEK	Swedish kronor
SOx	sulphur oxides
TSO	transmission system operator
TW	terawatt
VAT	value added tax
VAWT	vertical axis wind turbine
VOC	volatile organic compounds
WAsP	Wind Atlas Analysis and Application Program
ZVI	zone of visual impact

Glossary

AC Alternating current.
Acceptance People's attitude to wind power.
Acceptance, local Attitudes of people living close to wind turbines.
Aerodynamic brake Brake mechanism that reduces lift on blades by turning the blade tip or the entire blade.
Anemometer Instrument that measures wind speed.
Angle of attack The angle between the blade chord and the apparent wind direction.
Asynchronous generator Induction generator that runs with asynchronous rotational speed and is magnetized and governed by the grid frequency.
Availability (technical) The proportion of the time when the turbine is ready to operate.
Average wind speed Average of wind speeds registered during a specific time period.
Axial force Force applied to the rotor parallel with the wind direction (horizontal).
Battery charger Small wind turbine that charges a battery that is used for electric power supply in off-grid systems.
Bedplate The frame in the nacelle on which the gearbox, generator and other components are mounted. The frame is usually manufactured in cast iron or welded steel plates.
Betz's law Formulated by Alfred Betz and stating that an ideal wind turbine can utilize no more than 16/27 of the power in the wind.
Blade A wind turbine has one, two, three or more rotor blades. (Turbines do not have 'propellers' or wings.)
Blade area The product of the blade surface area and the number of blades (see *solidity*).

Blade form The form of the rotor blade: how the width is changed from blade root to blade tip and how it is twisted.

Blade profile Curved surface designed to create aerodynamic lift.

Blade root Part of rotor blade at the hub and the end that is attached to the hub.

Blade tip The outer end of the blade.

Boundary layer (internal) The boundary between two different layers of a fluid (like air) where the speed of the lower layer has been influenced by, for example, a change in surface roughness.

Breaker Circuit breaker that disconnects a wind turbine from the power grid.

Building permission Permission from an authority to build a house, wind turbine, etc.

Capacity factor The actual production of a power plant in relation to its production at full (nominal) power.

Capital cost Cost of amortizations and interest on loans.

Capital intensive A plant where the capital cost is high in relation to the operational costs. In the power sector hydropower and wind power are capital intensive, with high investment costs and low operational costs, since the fuel (water and wind) is free.

Comprehensive plan Municipal (or regional) plan on how land and sea areas should be utilized.

Conical tube tower Tube towers that narrow from the base to the top.

Constant rotational speed Rotor that rotates at a constant speed, independent of the wind speed.

Control system Computer system that controls the operation of a wind turbine.

Converter Component that converts AC to DC or vice versa.

Cooperative An association where, for example, members commonly own and operate a wind turbine.

Cube factor Factor that is used to calculate the energy content of the wind when only the average wind speed is known; important because the energy content is proportional to the cube of the wind speed.

Cut-in wind speed The wind speed at which a wind turbine starts to produce electric power.

Cut-out wind speed The wind speed at which a wind turbine is disconnected from the grid and stopped.

Darrieus turbine Vertical axis wind turbine with curved blades along the axis/tower; 'egg-beater'.

Data logger Equipment that collects and stores wind data.

dB, dB(A) See *decibel.*

DC Direct Current.

Decibel Unit to measure sound.

Direct drive generator Generator connected directly to the rotor without an intermediate gearbox.

Distribution grid The part of the power grid to which consumers' properties are connected.

Double wound generator Generator that can change number of poles and rotational speed; operates like two generators, one large and one small.

Downwind rotor Rotor at the back (leeside) of the tower.

Durability curve Graph that illustrates the durability of different wind speeds, etc.

Dutchman Yaw system that is driven by one or two wind wheels mounted perpendicular to the rotor.

EIA See *environment impact assessment.*

Eigenfrequency A frequency at which tower, blades or other components naturally vibrate. If these vibrations increase in strength (when, for example, the rotational speed of the rotor has the same frequency as the tower's eigenfrequency) the components can break.

Electric power grid Network of power lines for distribution of power from power plants to consumers.

Energy content The energy content of the wind at a specific site and height agl is signified by kWh/m^2/year.

Energy density See *power density.*

Energy rose Circular diagram that shows the distribution of the wind energy in different wind directions.

Environment impact assessment Public consultation process and report on the environment impact of a plant (such as a wind power project).

External costs Costs caused by power production that are not included in the power price, for example costs for health care and damage to the environment.

Feather To turn the rotor blades so that the wind can pass more easily and lift decreases.

Flicker Short (less than one second) variations in grid voltage.

Gearbox Mechanical equipment that changes the revolution speed of a rotating shaft.

Generator Machine that generates electric power.

Giromill Vertical axis wind turbine with two or more vertical blades.

Gondola See *nacelle*.

Grid benefit The benefit of local power production for local consumption; reduces losses in the grid.

Grid connection Connection of wind turbines to the power grid.

Grid integration Adaption of wind turbines to (the demands of) the power grid in a larger area, or vice versa.

Hill effect Increase of wind speed at the top of a hill.

Hub Centre of the rotor. The rotor blades are mounted to the hub and the hub is mounted to the main shaft. The hub is usually manufactured in cast iron.

Hub height The height from ground level to the centre of the rotor. The hub is a little higher than the tower height.

Infrasound Long-wave sound with a frequency < 20Hz that can't be perceived by the human ear.

Installed power Nominal power of a wind turbine; used to give a measure of the amount of wind power installed in a wind farm, region or country.

Intermittent energy source Energy source where the power output varies due to variations in the climate – wind speed for wind turbines, solar radiation for photovoltaic cells.

Isovent Isoline that shows the energy content of the wind on a wind resource map.

Laminar flow Airflow parallel to the horizontal plane with low turbulence.

Land breeze Wind from land to sea that is generated by temperature differences between land and sea.

Land lease Contract on lease of land (for example for a wind turbine).

Lattice tower Towers built using cross bars; can be used for wind turbines.

Leeward Direction that the wind blows to (opposite to windward).

Lift Force created on the upper side of a blade profile.

Main bearing Bearing mounted to the main shaft.

Main shaft Shaft (axis) connected to the rotor.

Maintenance costs Costs to maintain a wind turbine – servicing and repairs.

Mast, guyed Tower made of a mast guyed by steel cables.

Mechanical brake	Disc brake mounted on the main or secondary shaft between the gearbox and the generator; used as parking brake or emergency brake to stop the rotor.
Nacelle	The unit at the top of the tower that contains the gearbox, generator and other components.
NIMBY	'Not in my backyard'; expression for people who may be positive about wind power in general but not in their own vicinity, for example.
Noise	Annoying or undesired sound.
Nominal power	Nominal power of a wind turbine's generator.
Nominal wind speed	The wind speed at which a wind turbine reaches its nominal power.
Offshore	At sea.
Operation costs	Costs to operate a plant.
Opinion	An attitude to something (for example wind power) among a group of people.
Park efficiency	Production of a group of wind turbines as a proportion of the production they would have collectively if they didn't 'steal' wind from each other.
Pay-back time	The time it takes for incomes to cover investment costs.
Pitch angle	Angle between a blade's plane of rotation and the blade chord.
Pitch control	Power control by turning rotor blades on their axis.
PM-generator	Generator with permanent magnets.
Power, active	Useful electric power (measured in W); the power that is measured and the producer gets paid for.
Power, reactive	Useless electric power (measured in VAr – VoltAmpere reactive); created by phase displacement of current and voltage in alternating current.
Power coefficient	Share of the wind energy that a wind turbine rotor utilizes, signified by C_p and varying at different wind speeds. The theoretical maximum value is 0.59 (see also *Betz's law*).
Power curve	Graph that shows the relationship between wind speed and power from a wind turbine.
Power density	The average power of the wind at a specific site and height (measured in W/m^2).
Power electronics	Electronic components used for electric power.

Precautions	Measures taken to prevent, reduce or compensate for impacts on the environment.
Predominant wind direction	The most frequent wind direction.
Rayleigh distribution	Probability distribution that is used when the wind's frequency distribution is unknown.
Rectifier	Rectifies AC to DC.
Renewable energy sources	Energy sources that utilize resources that are not consumed, for example flowing water, wind and solar radiation.
Revolution speed	See *rotational speed.*
Ring generator	Multi-pole generator that runs with low rotational speed. Wind turbines with ring generators don't need a gearbox.
Rotational speed	Revolutions per minute, rpm.
Rotor hub	See *hub.*
Rotor plane	The plane that the rotor sweeps.
Rotor shaft	See *main shaft.*
Roughness class	A measure of surface roughness; the airflow's friction against the ground surface.
Roughness length	Another measure for surface roughness.
Savonius rotor	Type of vertical axis wind turbine.
Sea breeze	Wind from lakes or the sea towards the shore generated by temperature differences between ground and water.
Smoothing effect	The smoothing of the power output from wind turbines that are distributed over a geographical region.
Solidity	Blade area as a percentage of rotor swept area; a modern three-bladed turbine has a solidity of roughly 3 per cent.
Sound emission	Sound emitted from a sound source, for example a wind turbine.
Sound immission	Sound at a specific distance from a sound source.
Stall control	Power control used on turbines with fixed blade angle whereby eddies are generated when the wind speed increases above a certain limit.
Stand alone	Power system or wind turbine that operates isolated from the power grid; off-grid.
Survival wind speed	The wind speed that a wind turbine is designed to withstand without breaking down.
Teetering hub	Hub for two-bladed turbine, where the rotor can teeter a few degrees across the hub.

Tilt The angle of the main shaft towards the horizontal plane. Most turbines have a few degrees of tilt to prevent the rotor from sweeping too close to the tower and to move the centre of gravity closer to the tower.

Tip speed The speed of the blade tip of a wind turbine.

Tip speed ratio Relationship between the tip speed of the rotor blade and the undisturbed wind speed.

Torque The force that the rotor transmits to the main shaft and which makes it revolve.

Total height The total height of tower and rotor, i.e. hub height plus half the rotor diameter.

Transformer Equipment that increases the voltage level, for example from the 690V from a wind turbine to the 10kV in the power grid.

Turbulence Short and fast variations in wind speed.

Twist The difference between the blade angle at the blade root and that at the blade tip.

Ultrasound Short-wave sound with a frequency that can't be perceived by the human ear.

Upwind rotor Rotor on the front side of the tower.

Variable speed Wind turbines where the revolution speed increases with the wind speed (to keep the tip speed ratio constant).

Visual impact Impact (of wind turbines, for example) on the visual impression of a landscape.

Wake decay constant A constant used to calculate the wind wake's behaviour and extension on the leeside of a wind turbine rotor.

Weibull distribution Probability distribution that fits well to the frequency distribution of wind speed.

Wind climate The wind's long-term pattern in a country, region or site.

Wind–diesel system A stand-alone system where electric power is produced by wind turbines and diesel generators in a common system.

Wind farm Plant with several wind turbines.

Wind gradient Graph that shows how the wind speed will change with height above ground level.

Wind power cooperative See *cooperative.*

Wind profile See *wind gradient.*

Wind regime See *wind climate.*

Wind rose	Circular diagram that shows average wind speed, frequency or energy content for different wind directions.
Wind shear	A change of wind direction and/or speed with height agl.
Wind vane	Instrument that registers wind direction.
Wind wake	Area on the leeside of the rotor where wind speed and turbulence have been affected by the rotor.
Wind wheel	Rotor with many inclined blades that cover a large share of the rotor area; often used for wind pumps, since they exert high torque at low wind speeds.
Windward	The direction the wind comes from.
Yaw	Turning of the nacelle and rotor.
Yaw motor	Motor that turns the nacelle and the rotor towards the wind.
Yaw system	Control system and components used to turn the nacelle and rotor towards the wind.
Zone of visual impact	Area within which wind turbines can be seen.

Bibliography

This section includes details of all works cited in this book, as well as some additional publications used as background material but not directly referred to. A list of websites referred to in the book follows the main bibliography.

Ackermann, T. (ed) (2005) *Wind Power in Power Systems*, Wiley, Chichester, UK

Aldén, L. (2004) *Storbritanniens satsning på Offshore Vindkraft*, Gotland University, Visby, Sweden, available from www.cvi.se

AWS Truewind (2006) 'Wind forecasting eWind – proven, accurate, valuable', www.awstruewind.com/inner/services/windforecating/ewind/, accessed February 2006

Blomqvist, H. (1997) *Elkraftsystem*, Liber, Stockholm, Sweden

Bogren, J., Gustavsson, T. and Loman, G. (1999) *Klimatologi, Meteorologi*, Studentlitteratur, Lund, Sweden

Boverket (2002) *Förutsättningar för storskalig utbyggnad av vindkraft i havet, Vänern och fjällen*, VindGIS, Karlskrona, Sweden, available from www.boverket.se

Boverket (2003). *Planering och prövning av vindkraftsanläggningar*, Boverket, Karlskrona, Sweden, available from www.boverket.se

Burton, T., Sharpe, D., Jenkins, N. and Bossanyi, E. (2001) *Wind Energy Handbook*, Wiley, Chichester, UK

Claesson, P. (1989) *Vindkraft i Sverige – Teknik, Tillämpningar, Erfarenheter*, SERO, Köping, Sweden

Crown Estate (2003) 'Energy and telecoms: Offshore wind energy', available from www.crownestate.co.uk/34_offshore_wind_energy, accessed March 2006

Danish Wind Turbine Owners' Association (2001) 'Vindmöller og drivhuseffekten', www.dkvind.dk/fakta/, accessed March 2006

Danish Wind Turbine Owners' Association (2002) 'Hvem ejer vindmoellerne?', www.dkvind.dk/fakta/okonomi.htm, accessed March 2006

DENA (2005) 'Planning of the grid integration of wind energy in Germany, onshore and offshore up to the year 2020. Summary', Deutsche Energie-Agentur, www.dena.de

Drazga, B. (2001) 'Wind prediction models', *WindStats Newsletter*, vol 14, no 3, p5

EC (2001) 'Directive 2001/77/EC on the promotion of electricity produced from renewable energy sources in the internal market', http://europa.eu.int/comm/energy/res/legislation/, accessed December 2005

Eggersglüss, W. (2002) 'Das steht und das dreht sich', in *Ministerium für Finanzen und Energie des Landes Schleswig-Holstein. Stimmen zur Windenergie*, Ministerium für Finanzen und Energie des Landes Schleswig-Holstein, Kiel, Germany, pp8–10, available from http://landesregierung.schleswig-holstein.de

Elforsk (2005) *Driftuppföljning av Vindkraftverk, Årsrapport 2004*, www.vindenergi.org/vindstatistik/arsrapp2004.pdf

Elsam (2004) 'Life cycle assessment of offshore and onshore sited wind farms', Elsam Engineering report 186768 of March 2004, www.vestas.com/pdf/miljoe/, accessed January 2006

Enzensberger, N., Fichtner, W. and Rentz, O. (2003) 'Financing renewable energy projects via closed-end funds – a German case study', *Renewable Energy*, vol 28, pp2023–2036

EWEA (2003) *Wind Energy – The Facts*, report available for download at www.ewea.org/index.php?id=91&no_cache=1&sword_list[]=facts, accessed November 2006

EWEA (2005a) *Support Schemes for Renewable Energy*, report available from www.ewea.org/index.php?id=178, accessed November 2006

EWEA (2005b) *Large Scale Integration of Wind Energy in the European Power Supply*, report available from www.ewea.org/index.php?id=178, accessed November 2006

ExternE (2002) 'Externalities of energy', www.externe.info/publications

Focken, U., Lange, M. and Waldl, H.-P. (2001) *Previento – A Wind Power Prediction System with an Innovative Upscaling Algorithm*, EWEC, Copenhagen, Denmark

Freund, H. D. (2002) 'Einflüsse der Lufttrübung, der Sonnenausdehnung och der Flügelform auf dem Schattenwurf von Windenergieanlagen', *DEWI Magazine*, vol 20

Gasch, R. and Twele, J. (2002) *Wind Power Plants*, James & James, London

Gipe P. (1993) *Wind Power for Home and Business*, Chelsea Green Publishing Co., White River Junction, VT

Gipe, P. (1995) *Wind Energy Comes of Age*, Wiley, Chichester, UK

Gipe, P. (2004) *Wind Power*, James & James, London

Gipe, P. and Canter, B. (1997) *Glossary of Wind Energy Terms*, Vistoft Forlag, Knebel, Denmark

Gotlands Kommun (1999) *Vindkraft på södra Gotland, tillägg till översiktsplanen Vision Gotland 2010*, Stadsarkitektkontoret, Visby, Sweden

Greenpeace (2005) *Wind Force 12*, EWEA Publications, Brussels, Belgium

Hansen, M. O. L. (2000) *Aerodynamics of Wind Turbines*, James & James, London

Hansen, S. H. (2005) 'Wind energy in a spatial planning context', in *Proceedings from Wind Energy in the Baltic Sea Region*, www.cvi.se, accessed November 2005

Hedberg, P (2004). 'Vikande stöd för vindkraften', in *Ju mer vi är tillsammans*, SOM no 34, Gothenburg University, Gothenburg, Sweden

Hellström M (ed) (1998) *Vindkraft i Harmoni*, Energimyndigheten, Malmö, Sweden

Henderson, A. L., Morgan, C., Smith, B., Sorensen, H. C., Barthelmie, R. and Boesmans, B. (2002) 'Offshore wind power – A major new source of energy for Europe', *International Journal of Environment and Sustainable Development*, vol 1, no 4

Hills, R. L. (1996) *Power from Wind*, Cambridge University Press, Cambridge, UK

Holttinen, H. (2004) *The Impact of Large Scale Wind Power Production on the Nordic Electricity System*, VTT Publications 554, Helsinki, Finland, available at www.vtt.fi/renewables/windenergy/

Jacobsson, S. and Johnson, A. (2003) 'The development of a growth industry – The wind turbine industry in Germany, Holland and Sweden', in S. Metcalfe and U. Cantner (eds) *Change, Transformation and Development*, Physica-Verlag, Heidelberg, Germany

Krieg, R. (1997) *Vindenergikartering för Södra Sverige, del 1: Gotlands län*, Nutek, Stockholm, Sweden

Kullander, S., Rodhe, H., Harms-Ringdahl, M. and Hedberg, D. (2002) 'Okunnigt avveckla kärnkraften', *Dagens Nyheter*, 7 April, Stockholm, Sweden

Länsstyrelserna i Skåne (1996) *Lokalisering av vindkraftverk och radiomaster i Skåne*, Länsstyrelserna i Skåne, Lund, Sweden

Liik, O., Oidram, R. and Keel, M. (2003) *Estimation of Real Emissions Reduction Caused by Wind Generators*, International Energy Workshop, International Institute for Applied Systems Analysis (IIASA), available at www.iiasa.ac.at/Research/ECS/IEW2003/Papers/2003P_liik.pdf, accessed November 2005

Manwell, J. F., McGowan, J. G. and Roge, A. L. (2002) *Wind Energy Explained*, Wiley, Chichester, UK

Milborrow D. (1998) 'How much land does wind need?', *WindStats Newsletter*, vol 11, no 1

Milborrow D. (2002) 'External costs and the real truth', *Windpower Monthly*, vol 18, no 1, January

Miyamoto, C. (2000) *Possibility of Wind Power: Comparison of Sweden and Denmark*, International Institute for Industrial Environment Economics, Lund University, Lund, Sweden

Montgomerie, B. (1999) *Vindkraftaerodynamik*, Compendium, FFA, Stockholm, Sweden

NINA (Norsk Institutt for Naturforskning) (2006) 'Fire havörner drept av vindmöller på Smöla', www.nina.no, accessed February 2006

Nitsch, J., Krewitt, J., Nast, M. and Viebahn, P. (2004) 'Ökologisch optimirter Ausbau der Nutzung eneubarer Energien in Deutschland', in *Auftrag des Bundesministeriums für Umwelt, Naturschutz und Reaktorsicherheit*, Bundesministerium für Umwelt, Naturschutz und Reaktorsicherheit (BMU), Berlin, Germany, available from www.umwelt-ministerium.de

NWCC (2001) *Avian Collision with Wind Turbines: A Summary of Existing Studies and Comparisons to Other Sources of Avian Collision Mortality in the United States*, National Wind Coordinating Committee, Washington, DC

Ohlsson H. (1998) *Vindkraft. Tillgänglighet och Kostnader*, Elforsk rapport 98:32

Palu, I. (2005) 'Wind energy integration in Estonia', in *Proceedings from Wind Energy in the Baltic Sea Region*, www.cvi.se, accessed November 2005

Pasqualetti, M. J., Gipe, P. and Righter, R. W. (2002) *Wind Power in View*, Academic Press, London

Petersen, E., Mortensen, N., Landberg, L., Höjstrup, J. and Frank, H. (1997) *Wind Power Meteorology*, Risoe-I-1206(EN), Risoe National Laboratory, Roskilde, Denmark

Pettersson, J. (2005) 'Waterfowl and offshore windfarms – a study 1999–2003 in Kalmar Sound, Sweden', in *Proceedings from Wind Energy in the Baltic Sea Region*, www.cvi.se, accessed November 2005

Plehn, M. (2005) 'Suitability areas for wind turbines in Rostock region, northeast Germany', *Wind Energy in the Baltic Sea Region Conference Proceedings*, available from http://mainweb.hgo.se/projekt/cvi.nsf/index3?OpenFrameset, accessed November 2006

Quaschning, V. and Zehner, M. (2003) 'Rückenwind durch Simulationsprogramme', *Sonne Wind & Wärme*, March, www.volker-quaschning.de/downloads/sww03-2003.pdf

Redlinger, R. Y., Andersen, P. D. and Morthorst, P. E. (2002) *Wind Energy in the 21st Century. Economics, Policy, Technology and the Changing Electricity Industry*, Palgrave, Basingstoke, UK

Remmers, H. and Betke, K. (1998) 'Messung und Bewertung von tieffrequentem Schall', *Fortschritte der Akustik – DAGA*, Deutsche Gesellschaft für Akustik e.V., Oldenburg, Germany, pp472–473

Risoe National Laboratory (no date) 'Short-term forecasting fact sheet for Prediktor', www.risoe.dk/zephyr/publ/Zephyr_Prediktor_FactSheet.pdf, accessed November 2006

Sagrillo, M. (2003) 'Putting wind power's effect on birds in perspective', www.awea.org/faq/sagrillo, accessed March 2006

Smedman, A. (undated) *Varför Blåser Det?* Compendium, Uppsala University, Sweden

Söder, L. (1997) *Vindkraftens effektvärde, Elforsk Rapport*, vol 97, no 27, Elforsk, Stockholm, Sweden

SOU (1999) *Rätt plats för Vindkraft*, Fakta info direkt, Stockholm, Sweden

STEM (1999) *Att Köpa Vindkraftverk*, Energimyndigheten, Eskilstuna, Sweden

STEM (2001) *Vindkraftplanering i en kustkommun: Exemplet Tanum*, STEM, Eskilstuna, Sweden

Stiesdal, H. (2000) '25 Års teknologiudvikling for vindkraft – Og et forsigtigt bud på fremtiden', *Naturlig Energi*, July 2000

Swedish Energy Agency (2004) 'Renewable electricity is the future's electricity', STEM report ET 26:2004, www.stem.se/, accessed March 2006

Swedish Government (2002) *Samverkan för en trygg, effektiv och miljövänlig energiförsörjning*, Publication Prop. 2001/02, Swedish Government, Stockholm, Sweden, p143

Taylor Nelson Sofres (2003) *Attitudes and Knowledge of Renewable Energy amongst the General Public* (JN9419 and JN9385), TNS, London

Troen, I. and Petersen, L. E. (1989) *European Wind Atlas*, Risoe National Laboratory, Roskilde, Denmark

University of Oldenburg (2002) 'Wind power prediction system Previento', http://ehf.uni-oldenburg.de/wind/previento/english, accessed February 2006

Widing, A., Britse, G. and Wizelius, T. (2005) *Vindkraftens Miljöpåverkan – Fallstudie av Vindkraftverk i Boendemiljö*, CVI, Visby, Sweden

Wind Directions (2003) 'A summary of opinion surveys on wind power', in *Wind Directions*, September/October

Windpower Monthly (2006) 'Nuclear suddenly the competitor to beat', *Windpower Monthly*, January

Wizelius, T. (1992) *Vind, del 1*, Larson, Täby, Sweden

Wizelius, T. (1993) *Vind, del 2*, Larson, Täby, Sweden

Wizelius, T., Widing, A. and Britse, G. (2005) *Vindkraftens Miljöpåverkan – Utvärdering av Regelverk och Bedömningsmetoder*, CVI, Visby, Sweden

Wolsink, M. (2000) 'Wind power and the NIMBY-myth: Institutional capacity and the limited significance of public support', *Renewable Energy*, vol 21, no 1, pp49–64

World Wind Energy Association (2006) 'World wide wind energy boom', www.wwindea.org, accessed 26 October 2006

Yale University (2005) 'Survey on American attitudes on the environment – Key findings', Yale University School of Forestry and Environmental Studies, www.yale.edu/envirocenter, accessed November 2006

Web resources

ALWARE program by Ammonit	www.ammonit.de
ASTAE (The World Bank's Asia Alternative Energy Program)	www.worldbank.org/astae
AWS Truewind	www.awstruewind.com
Country Guardian (anti-wind farm site)	www.countryguardian.net
Danish Wind Industry Association	www.windpower.org
EMD International	www.emd.dk
Enercon GmbH	www.enercon.de
European Wind Energy Association	www.ewea.org
Greenius software by FHTW Berlin	www.f1.fhtw-berlin.de/studiengang/ut/downloads/greenius/
IEA Wind (International Energy Agency division)	www.ieawind.org
Prediktor (part of Zephyr family of software)	www.prediktor.dk/
Renewables for Sustainable Village Power	www.rsvp.nrel.gov/wind_resources.html
RETScreen International	www.retscreen.net
Siemens AG	www.powergeneration.siemens.com
WAsP (wind atlas analysis and application program)	www.wasp.dk
Wind Atlases of the World by Risoe National Laboratory	www.windatlas.dk
WindFarm program by Resoft	www.resoft.co.uk
WindFarmer program by Garrad Hassan and Partners	www.garradhassan.co.uk.
The Windicator	www.windpower-monthly.com
WinWind	www.winwind.fi

Index